DATA SCIENCE
AND ANALYTICS
WITH PYTHON

Chapman & Hall/CRC
Data Mining and Knowledge Discovery Series

SERIES EDITOR
Vipin Kumar
University of Minnesota
Department of Computer Science and Engineering
Minneapolis, Minnesota, U.S.A.

AIMS AND SCOPE

This series aims to capture new developments and applications in data mining and knowledge discovery, while summarizing the computational tools and techniques useful in data analysis. This series encourages the integration of mathematical, statistical, and computational methods and techniques through the publication of a broad range of textbooks, reference works, and handbooks. The inclusion of concrete examples and applications is highly encouraged. The scope of the series includes, but is not limited to, titles in the areas of data mining and knowledge discovery methods and applications, modeling, algorithms, theory and foundations, data and knowledge visualization, data mining systems and tools, and privacy and security issues.

PUBLISHED TITLES

ACCELERATING DISCOVERY: MINING UNSTRUCTURED INFORMATION FOR HYPOTHESIS GENERATION
Scott Spangler

ADVANCES IN MACHINE LEARNING AND DATA MINING FOR ASTRONOMY
Michael J. Way, Jeffrey D. Scargle, Kamal M. Ali, and Ashok N. Srivastava

BIOLOGICAL DATA MINING
Jake Y. Chen and Stefano Lonardi

COMPUTATIONAL BUSINESS ANALYTICS
Subrata Das

COMPUTATIONAL INTELLIGENT DATA ANALYSIS FOR SUSTAINABLE DEVELOPMENT
Ting Yu, Nitesh V. Chawla, and Simeon Simoff

COMPUTATIONAL METHODS OF FEATURE SELECTION
Huan Liu and Hiroshi Motoda

CONSTRAINED CLUSTERING: ADVANCES IN ALGORITHMS, THEORY, AND APPLICATIONS
Sugato Basu, Ian Davidson, and Kiri L. Wagstaff

CONTRAST DATA MINING: CONCEPTS, ALGORITHMS, AND APPLICATIONS
Guozhu Dong and James Bailey

DATA CLASSIFICATION: ALGORITHMS AND APPLICATIONS
Charu C. Aggarawal

DATA SCIENCE AND ANALYTICS WITH PYTHON

Jesús Rogel-Salazar

CRC Press
Taylor & Francis Group
Boca Raton London New York

CRC Press is an imprint of the
Taylor & Francis Group, an **informa** business

A CHAPMAN & HALL BOOK

CRC Press
Taylor & Francis Group
6000 Broken Sound Parkway NW, Suite 300
Boca Raton, FL 33487-2742

© 2017 by Taylor & Francis Group, LLC
CRC Press is an imprint of Taylor & Francis Group, an Informa business

No claim to original U.S. Government works

Printed in Great Britain by Ashford Colour Press Ltd.
Version Date: 20170517

International Standard Book Number-13: 978-1-498-74209-2 (Hardback)

Visit the Taylor & Francis Web site at
http://www.taylorandfrancis.com

and the CRC Press Web site at
http://www.crcpress.com

To A. J. Johnson and Prof. Bowman

Thanks to Alan M Turing for

opening up my mind

Contents

List of Figures

List of Tables

Preface

THIS BOOK IS THE RESULT of very interesting discussions, debates and dialogues with a large number of people at various levels of seniority, working at startups as well as long-established businesses, and in a variety of industries, from science to media to finance. The book is intended to be a companion to data analysts and budding data scientists that have some working experience with both programming and statistical modelling, but who have not necessarily delved into the wonders of data analytics and machine learning. The book uses Python[1] as a tool to implement and exploit some of the most common algorithms used in data science and data analytics today.

[1] Python Software Foundation (1995). Python reference manual. http://www.python.org

It is fair to say that there are a number of very useful tools and platforms available to the interested reader such as the excellent open source R project[2] or proprietary ones like SPSS® or SAS®. They are all highly recommended and they have their strengths (and weaknesses). However, given the experience I have been lucky to have had in implementing and explaining algorithms, I find Python to be a very malleable tool. This reminds me of a conversation with an

[2] R Core Team (2014). R: A language and environment for statistical computing. http://www.R-project.org

experienced analyst at a big consultancy firm who
mentioned that doing any machine learning or data science
related task in Python was impossible. I politely disagreed.
It is true though that there may be more suitable tools for
certain tasks, but it would be a truly Herculean labour to
present them all in one single volume. With that in mind,
the choice of using Python throughout this book suggested
itself: Python is a popular and versatile scripting and
object-oriented language, it is easy to use and has a large
active community of developers and enthusiasts, not to
mention the richness of the iPython/Jupyter Notebook, as
well as the fact that it has been used by both business and
academia for some time now.

We shall show in this book that
doing machine learning or data
science with Python is indeed
possible.

iPython/Jupyter Notebook is a
flexible web-based computational
environment that combines code,
text, mathematics and plots in
a single document. Visit `http:
//ipython.org/notebook.html`

The main purpose of the book is to present the reader with
some of the main concepts used in data science and
analytics using tools developed in Python such as
Scikit-learn[3], Pandas[4], Numpy[5] and others. The book is
intended to be a bridge to the data science and analytics
world for programmers and developers, as well as graduates
in scientific areas such as mathematics, physics,
computational biology and engineering, to name a few. In
my experience, the background and skills acquired by the
readers I have in mind are a great asset to have. However, in
many cases the bigger picture is somewhat blurred due to
the sharp specialisms required in their day-to-day activities.
This book thus serves as a guide to exploit those skills in the
data science and analytics arena. The book focusses on
showing the concepts and ideas behind popular algorithms
and their use, but it does not get into the details of their

[3] Pedregosa, F., G. Varoquaux,
A. Gramfort, V. Michel, et al.
(2011). Scikit-learn: Machine
learning in Python. *Journal of
Machine Learning Research 12*,
2825–2830
[4] McKinney, W. (2012). *Python
for Data Analysis: Data Wrangling
with Pandas, NumPy, and IPython.*
O'Reilly Media
[5] Scientific Computing Tools
for Python (2013). NumPy.
`http://www.numpy.org`

implementation in Python. It does, however, use open source implementations of those algorithms.

The examples contained in this volume have been tested in Python 3.5 under MacOS, Linux and Windows 7, and the code can be run with minimal changes in a Python 2 distribution. For reference, the versions of some of the packages used in the book are as follows:

- Python - 3.5.2

- Pandas - 0.19.1

- NumPy - 1.11.2

- Scikit-learn - 0.18

- StatsModels - 0.6.1

In particular I have chosen to use the Anaconda Python distribution[6] provided by Continuum Analytics as it offers installations in all of the three computer systems mentioned above, plus having the advantage of offering a rich ecosystem of libraries readily available directly from the distribution itself, and most importantly it is available to all. There are a few other ways of obtaining Python as well as other versions of the software: For instance directly from the Python Software Foundation, as well as distributions from Enthought Canopy, or from package managers such as Homebrew. Anaconda offers an easy environment to install and maintain the software, with minimum hassle for the user. I assume that the reader is working with the computer via scripts as well as interactively in a shell.

[6] Continuum Analytics (2014). Anaconda 2.1.0. https://store.continuum.io/cshop/anaconda/

Python Software Foundation https://www.python.org

Enthought Canopy https://www.enthought.com/products/epd/

Homebrew http://brew.sh

The book shows the use of computer code by enclosing it in a box as follows:

```
> 1 + 1 # Example of computer code

  2
```

We have made use of a diple (>) to denote the command line terminal prompt shown in the Python shell. Please note that the same commands can be used in the iPython interactive shell or iPython/Jupyter notebook, although the look and feel may be quite different. As you may have already noticed, the book uses margin notes, such as the one that appears to the right of this paragraph, to highlight certain areas or commands, as well as to provide some useful comments.

This is an example of the margin notes used throughout this book.

The book is organised in a way that individual chapters are sufficiently independent from each other so that the reader is comfortable using the contents as a reference rather than a textbook. Inevitably, there will be occasions where certain topics make reference to other parts of the book and I will point out when that may be the case. I would also like to take this opportunity to mention that the implementations presented are by no means the only or best way to do things. Programming is pretty similar to the creative process of writing: The fact that you have a set of words does not imply that we all write reports in a poetic manner. I would be delighted to hear from you all about the implementations and changes you make to the code presented here. Do get in touch!

Programming is a creative process, and as such there is more than one way to do things.

We start in Chapter 1 with a discussion of what data science and analytics are, from the point of view of the process and results obtained. We pay particular attention to the data exploration process as well as the data munging that needs to be carried out prior to the application of algorithms and analysis.

The data science workflow is discussed on Chapter 1.

In Chapter 2 we take the opportunity to remind us of some important features of the Python language. The aim is to revisit some important commands and instructions that provide the base for the rest of the book. This will also give us the opportunity to revise some commands and instructions used in later chapters.

A Python primer is given in Chapter 2.

In Chapter 3 we cover basic elements of machine learning, pattern recognition and artificial intelligence that underpin the algorithms and implementations we will use in the rest of the book.

Chapter 3 covers the basics of machine learning, pattern recognition and artificial intelligence.

By the time Chapter 4 is reached we will have the necessary foundations to implement regression analysis using Python via both StatsModels and Scikit-learn. The main points in the usage of generalised linear models for regression are covered in this chapter.

Chapter 4 covers various regression algorithms

In Chapter 5 we talk about clustering techniques, whereas Chapter 6 covers classification algorithms. These two chapters are central to the data science workflow: Clustering enables us to assign labels to our data in an unsupervised manner; in turn we can use these labels as targets in a classification algorithm.

Chapters 5 and 6 cover clustering and classification techniques, respectively.

In Chapter 7 we introduce the use of hierarchical clustering, decision trees and talk about ensemble techniques such as bagging and boosting. It is worth pointing out that ensemble techniques have become a common tool among data scientists and you are highly recommended to check this section out.

Chapter 7 deals with hierarchical clustering decision trees and ensemble techniques

Dimensionality reduction techniques are discussed in Chapter 8. There we will cover algorithms such as principal component analysis and singular value decomposition. As an application we will talk about recommendation systems.

Chapter 8 talks about dimensionality reduction.

Last but not least, in Chapter 9 we will cover the support vector machine algorithm and the all important Kernel trick in applications such as regression and classification.

Chapter 9 deals with support vector machines.

The book was made possible, as I mentioned before, thanks to discussions, presentations and exchanges with colleagues both in academia as well as in business. I am very grateful for their input and suggestions. I would also like to thank my editor at CRC Press, Randi Cohen, as well as the technical reviewers for their comments and suggestions. Finally, the encouragement that my family and friends have given me to take up yet another writing project has been invaluable. This goes to you all!

London, UK

Dr Jesús Rogel-Salazar
February 2017

Reader's Guide

THIS BOOK IS INTENDED TO be a companion to any jackalope data scientist from beginners to seasoned practitioners. The material covered here has been developed in the course of my interactions with colleagues and students and is presented in a systematic way that builds upon previous material presented.

Read Chapter 1 to understant the Jackalope reference.

I highly recommend reading the book in a linear manner. However, I realise that different readers may have different needs, therefore here is a guide that may help in reading and/or consulting this book:

- Managers and readers curious about Data Science:

  - Start by reading Chapter 1 where you will learn what Data Science is all about

  - Follow that by reading Chapter 3 where an introduction to machine learning awaits you

  - Make sure you understand those two chapters inside-out; they will help you to understand your jackalope data scientists.

- Beginners:

  - If you do not have a background in programming, start with Chapter 2, where a swift introduction to Python is presented

  - Follow that by reading Chapter 1 and Chapter 3 to understand more about what Data Science is and the principles of machine learning.

- Readers familiar with Python:

  - You can safely skip Chapter 2 and go directly to Chapter 4

- Seasoned readers may find it easier to navigate the book by themes or subjects

  - **Regression** is covered in Chapter 4, including:

    * Ordinary least squares

    * Multivariate regression

    * LASSO and Ridge regression

    * Support vector machines for regression are covered in Section 9.1.3

  - **Clustering**:

    * K-means is covered in Chapter 5

    * Hierarchical Clustering is covered in Section 7.1

  - **Classification** is generally covered in Chapter 6 including:

    * KNN

    * Logistic regression

    * Naïve Bayes

* Support vector machines for classification are covered in Section 9.1.4

- **Decision Trees and Ensemble Techniques** are discussed in Chapter 7

- **Recommendation Systems** are introduced in Section 8.4

- **Text manipulation** examples are provided in Section 6.4.2 where tweets are used as the main data source.

- **Image manipulation** examples are provided in Sections 8.2.1 and 8.3.1

About the Author

DR JESÚS ROGEL-SALAZAR IS a Lead Data Scientist with experience in the field working for companies such as AKQA, IBM Data Science Studio, Dow Jones and others. He is a visiting researcher at the Department of Physics at Imperial College London, UK and a a member of the School of Physics, Astronomy and Mathematics at the University of Hertfordshire, UK. He obtained his doctorate in Physics at Imperial College London for work on quantum atom optics and ultra-cold matter.

He has held a position as senior lecturer in mathematics as well as a consultant and data scientist in the financial industry since 2006. He is the author of the book *Essential Matlab and Octave*, also published with CRC Press. His interests include mathematical modelling, data science and optimisation in a wide range of applications including optics, quantum mechanics, data journalism and finance.

1

Trials and Tribulations of a Data Scientist

THE EVER INCREASING AVAILABILITY OF data requires
the use of tools that enable businesses and researchers to
draw conclusions and make decisions based on the evidence
provided by the data itself. From performing a regression
analysis to determining the relationship between data
features, or improving on recommendation systems used
in e-commerce, data science and analytics are used every
day by all of us. This book is intended to provide those
interested in data science and analytics a perspective into
the subject matter using Python, a popular programming
language available for various platforms and widely used
both in business and academia.

Data science and analytics is used
every day by all of us.

Python will be used throughout
the book, get well acquainted with
it!

In this chapter we will cover what data science is and how
it is related to various disciplines from mathematics to
business intelligence and from programming to design.
We will discuss the characteristics that make a good data
scientist and the composition of a data science team. We
will also provide an overview of the typical workflow in a

data science and analytics project and shall see the trials and tribulations in the work cycle of a data scientist.

1.1 Data? Science? Data Science!

THE USE OF DATA AS evidence in support for decision making is nothing new. You only have to take a look at the original meaning of the word *statistics* as the analysis and interpretation of information relating to states such as economic and demographic data. Nowadays, the word statistics is either understood as a branch of mathematics that deals with the collection, analysis, interpretation and presentation of data; or more colloquially as a fact or figure obtained from a study based on large quantities of data. Simply take a look at the news on any given day and you will surely get to hear about statistics, proportions and percentages, all in support (or not) of a new initiative, plan or recommendation. The power of data is all around us and we use it all the time.

Statistics was originally understood as the analysis and interpretation of information about states.

Now, what about the word science? Well, you may remember from your school days that science is a system that enables the organisation of knowledge, based on testable evidence and predictions. Notice that key word *evidence* mentioned there again.

Science is organised knowledge.

No surprises here so far, right? From a very simplified point of view, the scientific method makes use of data and their analysis to acquire, correct and integrate knowledge. Nonetheless, data science is not just simply the direct use

However, Data Science ≠ Data + Science

of statistics, or the systematisation of data. How shall we understand that much loved combination of the words *data* and *science*?

1.1.1 So, What Is Data Science?

DATA SCIENCE AND ANALYTICS ARE rapidly gaining prominence as some of the more sought after disciplines in academic and professional circles. In a nutshell, data science can be understood as the extraction of knowledge and insight from various sources of data, and the skills required to achieve this range from programming to design, and from mathematics to storytelling.

> Data science skills range from programming to design, and from mathematics to storytelling.

There is no doubt that the term *data science* is a true neologism of our time. The term has started being used and, to a certain extent, even abused. As we have mentioned before data science is rather more than the sum of data on the one hand and science on the other one, although it is inevitably related to both concepts.

> In the case of defining data science, the whole is indeed greater than the parts.

Currently, data science can be considered a budding field with applications in a wide range of areas and industries, as well as in academic research. It is fair to say that it is elusive to define this emerging field, and throughout this book we shall consider *data science and analytics* as a portmanteau for a number of overlapping tasks related to data - from collection, provision and preparation, analysis and visualisation, curation and storage - that exploit tools from empirical sciences, mathematics, business intelligence, machine learning and artificial intelligence. The aim of these

> In this book we will use a practical definition for *data science* as a combination of overlapping tasks related to data with the aim to derive actionable decisions.

tasks is to enable effective, pragmatic and most importantly actionable decisions.

The motivation for data science and analytics in deriving valuable insights from data is great, and widely welcomed by businesses. However, this is a very challenging task. Companies such as Google, Netflix and Amazon have demonstrated that careful storage and analysis of data delivers a very competitive edge. These days there are easier and cheaper ways to collect large amounts of data than ever before, and mobile is becoming a ubiquitous presence. This has allowed companies, particularly start-ups, to develop in-house data science capabilities.

Careful storage and analysis of data delivers a very competitive edge.

Typical examples of data science products are better explained by the questions they aim to answer; these questions are the drivers to the acquisition and selection of the appropriate data to be interrogated in order to provide insight into an area of interest. I am sure you can come up with a few of examples relevant to you, but here are some that come to mind:

Some examples of data science outputs make it easier to clarify what the discipline does. This list is by no means exhaustive.

- What product will sell better in conjunction with another popular product?

Market basket analysis

- Who will be declared Prime Minister (or President, or winner; depending on the flavour of the government system of interest) in the next general election?

Predictive analytics

- How can customers be encouraged to spend a longer time in an online portal?

E-commerce

- Are there any discernible patterns that allow us to

Clustering analysis and market segmentation

characterise different groups of sales agents, customers or businesses?

- What advertisement should be placed on what site?

Advertising and marketing

- Given the interests of a customer, what other products can be recommended to them?

Recommendation systems

- What are the latest developments and breakout reports in newspapers and social media that may affect the industry of interest?

Social media analysis

- Given someone's interests and hobbies, who may be suitable potential partners?

Online services

- How can we keep potentially sensitive information protected and react proactively to information we store?

Cybersecurity

- How can we distinguish valid, relevant documents such as emails (ham), from invalid, irrelevant ones (spam)?

Classification analysis

- How to determine if a retail transaction is valid or not?

Fraud prevention

- What is the demand for a particular service at a particular time or place?

Demand forecasting

These are not questions that decision-makers, businesses and industries, large and small, have recently started formulating. So, why the resurgence in seeking answers to them? The main answer is the availability of potentially useful data, big or small, together with the impact of technology, computer science and statistics in everyday life. Out of the ingredients mentioned above, accessible datasets may be the most important one since without them the insight provided by technology alone is rather limited. After all, the plural of anecdote is not data. Having said that, it is

The availability of large volumes of data has enabled data science to flourish.

The plural of *anecdote* is not *data*.

important to note that this does not mean that every single data science case to be tackled falls into the category of so-called *big data*, particularly when we take into account that the adjective *big* can be used in a relative manner. We shall expand on this point later on in Section 1.3.1.

One important thing to bear in mind about the outputs of data science and analytics is that in the vast majority of cases they do not uncover hidden patterns or relationships as if by magic, and in the case of predictive analytics they do not tell us exactly what will happen in the future. Instead, they enable us to forecast what may come. In other words, once we have carried out some modelling there is still a lot of work to do to make sense out of the results obtained, taking into account the constraints and assumptions in the model, as well as considering what an acceptable level of reliability is in each scenario.

Note that this book is about data science, not necessarily about big data.

Predictive analytics do not tell us the future; instead they allow us to forecast.

Similarly, there is the tacit prerequisite of having accurate, timely data that can be readily utilised to make sense out of the modelling results, and reflect the state-of-the-art in an application. It is therefore imperative that decision makers as well as IT and business stakeholders take time to understand the information that will be needed, as well as being prepared to realise that certain data may not be fit for their purpose. It is indeed disheartening to come to terms with the fact that some data may not have the necessary features to be used in building a prediction, for example. Nonetheless, it is better to realise that is the case at an early stage, rather than relying on unsuitable results to make important decisions that impact the business.

For data to be useful it should be available and it has to be timely.

Realising that data may not be fit for answering the questions at hand is a difficult but important thing to bear in mind.

Even if data science may not yet be considered a well-defined subject, the number of academic and training programmes being offered by universities and at various workplaces has seen a healthy increase. This is a natural result of the need that exists for well-informed, capable experts that we get to call *data scientists*. So what do data scientists do and what do they look like? It will all shall be uncovered.

The need for capable data scientists in industry has seen a healthy increase in recent times.

1.2   *The Data Scientist: A Modern Jackalope*

THE NEW TERM USED TO describe the person that deals with the seemingly disparate array of tasks described above may seem to be yet another, more fashionable way to describe a statistician or a business analyst. However, we can certainly agree that there is a gap between the latter two, and that the skills required by a data scientist involve aspects that include both statistics and a strong business acumen, but also foundations in computer science, mathematics, modelling and programming, not to mention good communication skills. A simplified diagram of these skills and their relationship is shown in Figure 1.1.

A data scientist requires the knowledge of mathematics and computer science, but also a good business background.

In that sense a data scientist role goes beyond the collection and reporting on data; it must involve looking at a business application or process from multiple vantage points and determining what the main questions and follow-ups are, as well as recommending the most appropriate ways to employ the data at hand.

The role of a data scientist goes beyond the collection and reporting on data.

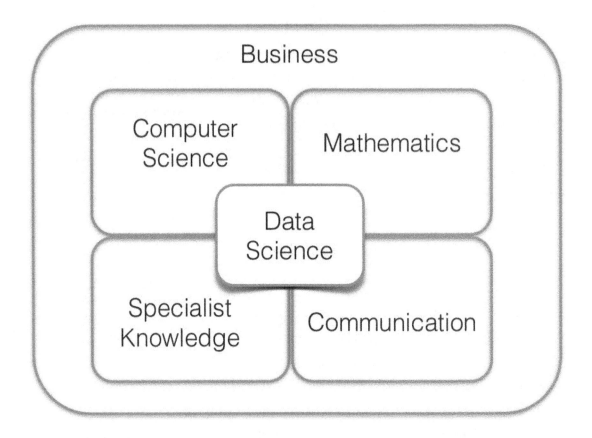

Figure 1.1: A simplified diagram of the skills needed in data science and their relationship.

In terms of characteristics, a data scientist has an inquisitive mind and is prepared to explore and ask questions, examine assumptions and analyse processes, test hypotheses and try out solutions and, based on evidence, communicate informed conclusions, recommendations and caveats to stakeholders and decision makers.

Not only does a data scientist need an inquisitive mind, but also good communication skills.

In other words, a data scientist is a true new Renaissance woman or man. No wonder that despite being branded the sexiest job of the 21st century, as well as the increasing demand for these individuals, it is hard to find people with the right skills to fill in these roles. This has lead to branding data scientists as Unicorns.

No wonder data scientists have been dubbed Unicorns.

To a certain extent, the symbolism of a Unicorn as a creature that is beautiful, mysterious and difficult to tame or even capture may be applicable to describe a data scientist. However, in my opinion, it may not be totally appropriate given the fact that, as majestic as Unicorns can be, they are way too common as far as popular culture goes.

Branding data scientists as Unicorns is a result of the quixotic expectations businesses and industries have and thus is not appropriate.

The shortage that businesses experience when trying to attract data scientists is more likely due to the fact that they have created internal expectations for the role and that no single individual can fulfil, thus appealing to the magical nature of a common mythical beast. They have created their idea of the *Data Scientist Unicorn*, and unfortunately the fascination prevails.

To tackle the prevailing image, I am convinced that the use of a new symbol is needed. And a silly one at that! There is an allegory I usually propose to colleagues and those that talk about the *data science Unicorn*. It seems to me to be a more appropriate one than the existing image: It is still another mythical creature, less common perhaps than the Unicorn, but more importantly with some faint fact about its actual existence: a *Jackalope*. You can see an artistic rendition of a couple of Jackalopes in Figure 1.2.

A new allegory is needed to tackle the unrealistic image of a data scientist.

The *Jackalope* is the one we propose.

A Jackalope is said to be a strange beast that looks like a jackrabbit with a pair of stag horns. It is described to be a shy but clever and cunning animal, and if threatened it can be dangerous. If you are ever in the Mountain West in

A Jackalope is a mythical being similar to a jackrabbit with a pair of stag horns.

Figure 1.2: Jackalopes are mythical animals resembling a jackrabbit with antlers.

the United States you may stumble into Jackalope heads mounted as trophies; but of course that is not the only place where Jackalopes are endemic; there are tales of the *Hasenbock* in Austria[1] or you may hear the Huichol stories in Mexico about how the *tátSu* (rabbit) lost its antlers to the *kaukamali* (deer)[2].

No need to explain that a Jackalope is indeed an imaginary, mythical being, much like the Unicorn, but it seems to be a better metaphor for the data scientist. We can argue that it is

[1] Toelken, B. (2013). *The Dynamics of Folklore*. University Press of Colorado

[2] Zingg, R., J. Fikes, P. Weigand, and C. de Weigand (2004). *Huichol Mythology*. University of Arizona Press

rather difficult at best, and impossible at worst, getting hold of a single individual that is able to be an all rounded ninja programmer, with vast expertise in mathematics, statistics and probability, plus knowledge of computer science and well-versed in business. This offers no solution to businesses interested in getting the benefit of exploiting the available data.

It is indeed difficult to get hold of a Unicorn.

Well, if you cannot get them in the wild, make them up from various parts - in the best style of Dr Frankenstein and his monster - and that is where the image of the Jackalope comes handy. In 1932 Douglas Herrick did indeed put together his creation when he stuck a pair of deer horns on a dead jackrabbit and mounted it as a trophy[3]. The rest is history, as the Converse County city of Douglas, Wyoming became the Jackalope capital of the United States.

If you cannot get data scientists in the wild, make them up.

[3] Martin, D. (2003, Jan 19th). Douglas Herrick, 82, Dies; Father of West's Jackalope. *The New York Times*

Furthermore, you do not have to get a fake hunter's trophy to see a Jackalope. As I mentioned before, there is faint fact to the existence of horned rabbits. That is definitely more than one can say about a one-horned horses. This is thanks to the existence of a virus, the cottontail rabbit papilloma virus (CRPV), which makes infected rabbits grow bone-like structures in their skulls[4]. The virus was discovered in the 1930s by Richard E. Shope and was the first example a cancer caused by a virus.

Plus, there is a faint fact to the existence of horned rabbits.

[4] Zimmer, C. (2012). *Rabbits with Horns and Other Astounding Viruses.* Chicago Shorts. University of Chicago Press

The use of this allegory is proposed to show how silly it is to simply employ wishful thinking in the pursuit of exploiting data and hoping that a single individual will come to the rescue. What I am trying to say is that one should think optimistically about the prospect of finding capable data scientists if we are prepared to be realistic about distinguishing mythological aspirations from messy reality.

It is possible to find capable data scientists if we are prepared to be realistic about our expectations.

What I propose is that the best way to tackle the data science needs of a business - a startup or a large conglomerate - is to put together a rangale of jackalope data scientists, than daydreaming of a bliss of non-existant Unicorns. After all, there are indeed better chances of seeing a Jackalope-like animal than a Unicorn, right?

I propose therefore to put together a rangale of jackalope data scientists.

The next question is thus related to how the rangale of data scientists should be put together, what roles they should have and what resources to provide them with. These points are perhaps not easy to answer, as they depend to a large extent on the area in a business where the insight is being sought, and for what purpose (see Section 1.4). Nonetheless, there are some general guidelines that can be taken into account when tackling the data scientist conundrum.

Not only is it important to know what qualities a data scientist should have, but also what role they are expected to play and what tools they will use to do their jobs.

1.2.1   *Characteristics of a Data Scientist and a Data Science Team*

IT SEEMS THAT EVERYONE LOVES, or would love to have, a data scientist, and as we have seen, the wishful list of desired characteristics makes it more difficult for businesses to choose among otherwise capable candidates.

The analogy that comes to mind is that of the everlasting dating puzzle where everyone is waiting for Princess or

Everyone would like to have their own data scientist and knowing what is important for the business needs is a major aspect to consider.

Prince Charming, but is unable to find "the one". For a data scientist to be considered "the one" the skills required include those discussed in the previous section and summarised in Figure 1.1.

Let us pause for a moment before we tackle the subject at hand and consider what the purpose of the data science team is or will be. This is a crucial step in building that team as these objectives will help identify the important traits that the data scientists are expected to have. Furthermore, having a clear idea of how they will fit in the organisation and what problems they are expected to solve will aid in defining the size of the team and the type of expertise needed. It is not uncommon to hear of organisations that are interested in riding the data science wave, but do not have a clear goal regarding the purpose of their data science journey.

Having a clear idea of how a potential data scientist will fit in the organisation and what they will work on is important.

With the objective of the data science team in mind, it becomes much easier to decide what is relevant in a particular case. In general, what makes a good data scientist is a linear combination of some of the following traits:

- Curiosity

- Grasp of machine learning

- Data product building and management

- Effective communication of data insights

Some important traits in a data scientist.

- Programming and data visualisation abilities

- Knowledge of statistics and probability (other mathematical areas are welcome)

- Healthy skepticism, in the scientific tradition: Carry out experiments, test hypotheses, etc

The important thing to realise here is that the linear combination of the features mentioned above do not necessarily have to be equally weighted, and that is the main reason for the persistance of the Unicorn fallacy we have been discussing. Should your data scientists lack some more developed branches in their antlers, all you need to do is give them a helping hand and provide them with colleagues that will help in developing those skills, but more importantly cover the gap in those desirable features. In other words, much like Mr Herrick, put together your very own Jackalope team with people who have a broad-range of generalist interests, but a deep expertise in a certain area or two.

The features mentioned do not have to be combined in equal measures.

The sensible thing to do is to start with a solid core and not let the list above let you get carried away. In other words, setting the foundations of the data science team is similar to having strong foundations in a building; without them the whole tower may collapse in an instant. Furthermore, use this core to your advantage and bank some of the easy wins to start with. The three pillars in this data science triumvirate I am referring to may include, with variations in the titles, the following main roles:

Start a data science team with a solid core, perhaps made out of more than one person.

- Data Science Project Manager

- Lead or Principal Data Scientist

- Data Architect

The data science triumvirate.

Having a person that is able and experienced in managing technical teams is an important role to have in the mix. The main idea is to cover the fact that many a data scientist is far more interested in tackling questions and problems head on, rather than dealing with managing a project from end to end. One way to help them deliver is to have a knowledgeable individual that is able, on the one hand, to keep track of how projects are going, attend meetings and manage relationships. On the other hand, they should have a general understanding of techniques, algorithms and technology to be able to liaise with the team effectively. The project manager does not have to be a ninja programmer, but should be able to understand what the rest of the team are working on and the challenges they may be facing.

First, a Data Science Project Manager is needed.

The second figure in the triumvirate is that of the principal data scientist. Not only is it necessary to have a good project manager, but also have someone with a strong background in a quantitative field: Physics, mathematics, computer science, etc. Ideally the academic credentials this person would speak for themselves. In terms of programming, this person may not be a developer in the full sense of the word, however, they should have a firm background in coding and solving problems with the use of technology. An important ingredient of the role is to be able to act as an advisor or guide to other data scientists and analysts in the team.

Followed by a Lead Data Scientist.

The third pillar in the team is the data architect, who will provide expertise in terms of data structures, databases, software engineering and computational capability. It is important for the data architect to be able to disentangle the data resources that the business may (or may not) have, and be able to use their expertise to assess what data is available, when it is available, and manage the constraints that the business, regulation and security impose on the workflow. Ideally, the data architect would be interested in quantitative topics, but most importantly their programming skills must be spot on. Note that the data architect will use the same technology that the data scientists employ in their day-to-day activities.

And finally a Data Architect completes the trio.

Finally, there are four aspects that are important to remember when considering putting together that data science team. First, consider who the main stakeholders of the data science team are, and clarify the lines of reporting. Remember that everyone wants their own data scientist, and confusing or conflicting messages can lead to undesired results.

Clear reporting lines are also important.

Second, for data scientists to be able to work independently and (more importantly) productively it is important for them to be able to navigate the stack entirely. This enables extracting relevant data with appropriate tools (see Section 1.3). A data science team without strong IT skills or engineering support will have a hard time doing the job they do best.

Having appropriate tools to work with is paramount.

Third, once data has been identified for tackling a problem, proper interpretation is not necessarily easy, and misrepresentation of the results can be very damaging. It is not uncommon to see the use of tools such as machine learning algorithms to be seen as a black box; in practice,

It is necessary to have appropriate expertise to interpret and rework results.

knowing the capabilities, limitations and trade-offs requires experience.

Fourth, have the product always in mind: Not only is it important to have the right IT and statistics/machine learning skills, but also the team has to have a clear idea of the final product of their efforts, as well as their target audience. You may be able to come up with the most amazing models and results, but they may not be of much use if the product is of no interest to stakeholders or if the data scientist fails to communicate the results to them.

Also have a clear idea of the final product and communicate results clearly.

Consider as well the tools used to present results; in other words, there may be technology out there that lets the data scientist dazzle his/her target audience, but if that audience is not able to even access the technology, then you have lost the battle before starting.

Appropriate technology for presenting and delivering results is also important.

A point in case in my experience is the use of great JavaScript libraries such as D3. I am an advocate for their use as they can be effective and even great fun to use. However, they only work on "modern" browsers and unfortunately a large number of institutions out there only support old browsers unsuitable to render the created assets. This becomes a relevant point when considering the deployment of solutions (dashboards, reports, etc.).

1.3 Data Science Tools

WITH OUR NEWLY ACQUIRED DATA science team and the individual high-calibre data scientists and analysts that compose it, we are able to keep abreast of the the latest developments in the field of analytics and data science, and are able to extract actionable insights from our data. However, not only do we need to be flexible, agile and expert, we are also required to have the right tools and infrastructure to enable the team to fulfil the objectives agreed with the team sponsors. To that end, there are a number of considerations that we would need to think about in helping the team decide on the tools needed as well as some other points such as:

The tools chosen need to enable us to be flexible agile and expert.

- Regulatory and security requirements of hosting and manipulating the data

- Locations of data sources - and related subjects such as whether we would need/have immediate access to them, or would get them in batches for upload

Some considerations when choosing appropriate tools.

- Responsiveness requirements for queries - e.g. Real-time v Fixed Reporting

- Volume of queries/searches to be run

- Format of the data source

- Quality of the data

The security consideration above is usually a big question for any business that requires their data to be in a particular jurisdiction and does not plan to create their own cloud service. For instance, Google to date will not guarantee that data will stay in Europe, for example.

Security of the data is very important.

Data science and analytics is all about data, statistical analysis and modelling. It is therefore important to have the technology that enables those functions. A data warehouse, ETL software, statistical, modelling and data-mining tools are necessary. Similarly, an appropriate hardware and network environment are required (perhaps even in the cloud).

A data warehouse, ETL software, statistical, modelling and data-mining tools are necessary.

The technologies used in the analytics arena have evolved at a fast pace in the last few years, and a number of open source projects, with lots of support have emerged, for instance:

- *Data Framework*: **MapReduce, BigQuery, Hadoop, Spark.** Hadoop is probably the most widely deployed (if sometimes under-utilised) framework to process data. Hadoop is an open source implementation of the MapReduce programming model from Google. Other technologies are aimed at processing streaming data, such as S4 and Storm. BigQuery (by Google) is a web service that enables interactive analysis of massive datasets and can be used in conjunction with MapReduce. Enterprise versions of Hadoop are available from vendors such as HortonWorks. More recently the use of Spark has captivated the imagination of the big data connoisseurs

Data framework technologies

- *Streaming data collection*: **Kafka, Flume, Scribe.** The models may be different but the aim is similar: Collect data from many sources, aggregate it and feed it to a database, or a system like Hadoop, or other clients

Streaming data collection technologies

- *Job scheduling*: **Azkaban** and **Oozie** manage and coordinate complex data flows

Job scheduling technologies

- *Big Data Query languages*: **Pig** and **Hive** are languages for querying large non-relational datastores. Big data

Big data query languages

frameworks such as MapReduce and Hadoop can be made more "user friendly" with them. Hive is very similar to SQL. Pig is a data-oriented scripting language

- *Data stores*: **Voldemort, Cassandra, Neo4j** and **HBase**. These are data stores designed for good performance on very large datasets

Data stores

1.3.1   *Open Source Tools*

THE MODEL OF DEVELOPING TOOLS whose source code is made available for contribution has shifted the environment for their deployment both in small and large enterprises. The collaborative nature of the various projects provides a pool of knowledge and quality assurance that is difficult to beat. A rich and wide set of tools in the open source domain has contributed to the expansion of data science. They include tools that process large datasets as well as data visualisation, together with prototyping tools:

There are many open source tools that can be readily used in the data science workflow.

- **Python**: Data manipulation, prototyping, scripting, and the main focus in this book

We will be using Python in this book.

- **Apache Hadoop**: Framework for processing big data

- **Apache Mahout**: Scalable machine-learning algorithms for Hadoop

- **Spark**: Cluster-computing framework for data analytics

- The **R** Project for Statistical Computing: Data manipulation and graphing

R is a noteworthy software package widely used by the data science community.

- **Julia**: High-performance technical computing

- **GitHub, Subversion**: Software and model management tools

- **Ruby, Perl, OpenRefine**: Prototyping and production scripting languages

As mentioned above, Hadoop is rapidly becoming ubiquitous for processing massive datasets. The framework is scalable for distributed data processing, but as remarked in Section 1.1.1, in my view not all data science problems require big data processing. The Hadoop "hype" has caused many organisations to deploy MapReduce-like systems that are effectively used to dump data - without a big picture of the information management strategic plan or without understanding how all the pieces of a data analytics environment fit together.

Not all data science is about big data.

R is seen as the programming language for statistical computing. It is not characterised by the beauty of its code, but the results are great. The number of packages that is available in the R repository (CRAN) makes it very flexible.

The use of scripting languages such as **Python** provide a professional platform for application development and deployment. It is very well suited for prototyping and testing new ideas. Furthermore it supports various data storage and communication formats, such as XML and JSON, plus there is a large number of open source libraries for scientific computing and machine learning.

In recent times, Python has seen a resurgence thanks to the data science scene.

Python has a number of very useful libraries such as SciPy, NumPy and Scikit-learn. SciPy extends Python into the

domain of scientific programming. It supports various
functions, including parallel programming tools, integration,
ordinary differential equation solvers, and even extensions
for including C/C++ code within Python code. Scikit-learn
is a Python-based machine learning package including
many algorithms for supervised learning (support for vector
machines, naïve Bayes), unsupervised learning (clustering
algorithms), and other algorithms for dataset manipulation.
It is for these reasons that we will use Python in the rest of
this book.

Python is a well supported
language with a wide variety of
modules and libraries.

1.4   From Data to Insight: the Data Science Workflow

AS WE HAVE SEEN ABOVE, the role of a data scientists is
an interesting one, and at times a challenging one. Not
only do we need the right combination of skills (either in a
team or an individual), but also the right tools and business
questions. In this section we will address the steps that a
data science project may follow. It is important to emphasise
that although we may categorise and separate the various
steps, the workflow is not necessarily a linear one as we
shall see.

The workflow of the data science
expert is worth discussing too.

With our newly acquired data science team and the right
combination of skills, we are ready to tackle out first
assignment, and it is now when key measurements of
success for the project should be identified. Furthermore, it
is important to realise from the start that in the vast majority
of cases there will not be a unique, final answer. It is thus
better to frame the problem as an iterative process where a

The data science process is
iterative.

better solution is reached on each iteration. The various
steps in the data science workflow include:

- Question identification

- Data acquisition

- Data munging

- Model construction

- Representation

- Interaction

The steps in the data science
workflow

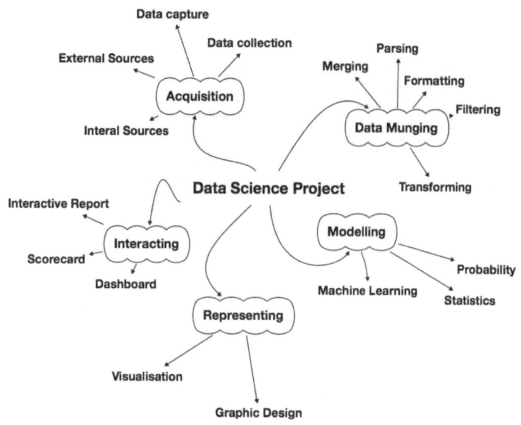

Figure 1.3: The various steps
involved in the data science
workflow.

The fact that they have been listed in that order does not
mean that they have to be followed one after the other. In
some cases you may start, for example, with an internal
dataset and immediately create some plots even before
cleaning the data. Also, once you have started the project,
you may move between steps in every iteration. Figure 1.3
shows a diagram of the steps mentioned above, note that
they do not necessarily follow one another in the order
listed above.

The workflow outlined above
is not necessarily followed in
sequence.

1.4.1   Identify the Question

THERE IS NO SUCH A thing as insight without a clear and
concise question, as well as having a way to measure the
success or failure of the answer obtained after running your
favourite machine learning algorithms. It is not a matter
of getting a dataset and simply massaging it and creating
some plots. On the contrary, let the questions point out the
potential datasets that may be useful in answering them and
to what extent.

Without a clear question, there is
no insight.

Another important thing to bear in mind is the fact that
although we may encapsulate a problem in a single
question, in many cases it is much easier to break it down
into smaller parts that can be tackled in a more
straightforward manner. Furthermore, at every iteration
there may be more, smaller or larger, follow-up questions
that will also require answers. Remember that it is an
iterative process!

Breaking down the problem into
smaller questions is useful.

1.4.2   *Acquire Data*

ONCE YOU HAVE A PROBLEM to tacke, the first thing that
needs doing is figuring out if you or your organisation has
the appropriate data that may be used to answer the
question. If the answer is no, you will need to find
appropriate sources for suitable data externally - web, social
media, government, repositories, vendors, etc. Even in the
case where the data is available internally, the data may be
in locations that are hard to access due to technology, or
even for regulatory and security reasons.

Identify appropriate sources of
suitable and useful data.

1.4.3   *Data Munging*

IF THERE IS NO INSIGHT without a question, then there
is no data without data munging. Munging, or wrangling
data is actually the most time-consuming task in the data
science workflow. According to the New York Times' Steve
Lohr data scientists may spend anything between 50 and 80
percent of their time doing "data janitor work"[5] and I can
definitely attest to that.

There is no data without data
munging.

[5] Lohr, S. (2014, Aug 17th). For
Big-Data Scientists, 'Janitor Work'
Is Key Hurdle to Insights. *The New
York Times*

Data preparation is key to the extraction of valuable insight
and although some may prefer to concentrate only on the
much more fun modelling part, the fact that you get to
know your dataset inside out while munging it implies that
any new or follow-up questions can probably be attained
with less effort.

1.4.4   Modelling and Evaluation

HAVING A CLEAN DATASET TO feed to a machine learning or
statistical model is a good start. Nonetheless, the question
remains regarding what the most appropriate algorithm
to use is. A partial answer to that question is that the best
algorithm depends on the type of data you have, as well
as its completness. It also depends on the question you
decided to tackle. Once the model has been run through the
so-called *training* dataset the next thing to do is to evaluate
how effective and accurate the model is against the *testing*
dataset and decide if the model is suitable for deployment.

Every model needs to be
evaluated.

For more in training and testing
datasets, see Section 3.11.

1.4.5   Representation and Interaction

THEY SAY THAT A PICTURE is worth a thousand words, and
it does stand to reason that the vast majority of us get
further information from a couple of well presented plots
than by looking at rows and rows of data. Data visualisation
is more of an art than a science, and much has been written
(or drawn) about by brilliant designers and data journalists.
Simply take a look at the great images produced by David
McCandless in his book *Information is Beautiful*[6] or the
visualisations produced by Manuel Lima in *Visual
Complexity*[7].

Data visualisation is more of
an art than a science, but an
important one nonetheless.

[6] McCandless, D. (2009). *Information
is Beautiful*. Collins

[7] Lima, M. (2011). *Visual Complexity:
Mapping Patterns of Information*.
Princeton Architectural Press

You do not have to produce such artistic beauties, but make
sure that the data representation that you decide to use is
accurate, simple, and provides clarification to the story you
want to communicate. In some cases, there is the possibility

of telling the story in a way that the reader/viewer is able to interact with the data representation, for instance in the form of dashboards, reports or interactive plots. These may be fun to use, but the same principles of accuracy, simplicity and clarity do apply.

1.4.6   Data Science: an Iterative Process

THE SIMPLE FACT THAT A machine learning model has been run on clean data does not mean that the work of the data scientist is done and dusted. On the contrary, the effectiveness of the model needs to be carefully monitored as the outcome depends on the data that is fed to them. A simple case of *garbage-in-garbage-out*. Similarly, any new data with a variety of new features may degrade the accuracy of the initial model, and thus it becomes necessary to adjust parameters or acquire new data.

The effectiveness of a model needs to be monitored.

Furthermore, even if there were such a thing as a never-changing model, the fact is that new and follow-up questions arise naturally from the data science process. This means that the workflow is better to be treated as an upward spiral where constant iterations provide improvement and new insights.

Think of this process as an upward spiral.

I would like to close this chapter with a few questions that the data scientist team and their stakeholders should always have in mind:

- What data was used and why?

- Where was the data acquired from and who owns it?

- Was the entire dataset used? Is a sample representative of the entire population?

- Were there any outliers? Have they been considered in the analysis?

- What assumptions were made when applying the model/algorithm? Are they easily relaxed/strengthened?

- What does the result of the model mean to the process/business/product?

Some questions to always bear in mind during any data science project.

1.5 Summary

IN THIS CHAPTER WE ADDRESSED some crucial aspects that will enable us to implement and acquire that elusive mythical being called a data scientist. We have provided a working definition for the term *data science* and have described how it is a rapidly evolving multi-disciplinary field encompassing areas in mathematics, computer science, statistics and business expertise.

We mentioned a few examples of data science products and have seen that the main motivation for data science and analytics is deriving valuable and actionable insights from data. Our discussion pointed out that in order to fulfil that motivation, a data scientist with the appropriate skills is needed. Unfortunately, understanding about this subject has created expectations that point at the data scientist role as one akin to the mythical Unicorn. We argued that a better understanding of the role indicates that we should perhaps use a different allegory: A Jackalope. It is still a mythical being, but the fact that you can put one together out of different parts (as in a team), or the hint that there may be a scientific explanation for their potential existence provides a more hopeful panorama for many businesses interested in exploiting their data.

We continued our discussion with the three main pillars that would sustain a productive data science team, namely a data science project manager, a lead or principal data scientist and a data architect. We saw that not only is the team composition important, but also the tooling provided to carry out their tasks in a productive manner. We concluded this chapter by outlining the steps in a simplified data science workflow and explained their importance as part of what should be seen as an iterative process.

In the next chapter we will provide a brief refresher of some important concepts of using Python. This will enable us to have a point of reference for the rest of the book where Python will be used to implement a variety of algorithms that form part of the Jackalope's bag of tricks.

2

Python: For Something Completely Different

THERE IS NO SHORTAGE OF programming languages and
paradigms. With that in mind it may seem surprising that
what started up as a "hobby programming project" by
Guido van Rossum in 1989 has taken on a life of its own,
with a very active community, used in a wide variety of
applications. Its success is perhaps due to the compactness
of its code, or the fact that is open source, or even the
variety of toolsets. Whatever the reason, Python is a great
tool to have under your jackalope data scientist belt.

Python started up as a hobby programming project.

Python is not named so after the nonvenomous snakes
whose family includes some of the largest in the World.
Instead it is named after the famous British comedy troupe
Monty Python, and it was created to be appealing to Unix/C
coders. Today, Python's emphasis on code readability and
expressive syntax has made it into a general-purpose, high-
level, object-oriented programming language available
in multiple platforms and with a plethora of supporting
packages and modules.

We have to thank Monty Python for the name of this flexible programming language.

```
> import this


The Zen of Python, by Tim Peters
Beautiful is better than ugly.
Explicit is better than implicit.
Simple is better than complex.
Complex is better than complicated.
Flat is better than nested.
Sparse is better than dense.
Readability counts.
Special cases aren't special enough to break
the rules.
Although practicality beats purity.
Errors should never pass silently.
Unless explicitly silenced.
In the face of ambiguity, refuse the temptation
to guess.
There should be one -- and preferably only one --
obvious way to do it.
Although that way may not be obvious at first
unless you're Dutch.
Now is better than never.
Although never is often better than *right* now.
If the implementation is hard to explain,
it's a bad idea.
If the implementation is easy to explain,
it may be a good idea.
Namespaces are one honking great idea
-- let's do more of those!
```

"Pythonic" coding style guidelines are followed by so-called Pythonistas.

Perhaps some of the most defining features of the language include the use of indentation for grouping statements, together with was has become to be known as the "pythonic" coding style, i.e. style guidelines and idioms used by veteran Python programmers, aka Pythonistas. These guiding principles can be read in any Python installation by importing the this module. These aphorisms, compiled by the long time pythonista Tim Peters, are shown in the code listing printed above.

The use of indentation is an important feature of Python.

This book assumes some knowledge of general programming principles and a level of familiarity with Python. Nonetheless, in this chapter we will briefly review some of the concepts and idiosyncrasies of the language that will be used throughout the book. If you are a seasoned Python programmer, you may want to skip this chapter and move on to the next one. If, however, you are interested in a refresher, then go ahead and read the following pages, you might find something completely different. For those of you that are beginning their Python journey, this chapter may whet your appetite to learn more. There are plenty of resources to quench your thirst[1,2].

We will assume a certain familiarity with programming principles and with Python in particular.

[1] Downey, A. (2012). *Think Python*. O'Reilly Media
[2] Langtangen, H. (2014). *A Primer on Scientific Programming with Python*. Texts in Computational Science and Engineering. Springer Berlin Heidelberg

2.1   *Why Python? Why not?!*

WE MENTIONED ABOVE THAT PYTHON emphasises code readability, and this in turn has an impact on productivity: Not only is a data scientist able to create scripts to be executed as a batch, but also is able to start an interactive console (iPython shell for instance) or notebook (iPython/Jupyter notebook - see Section 2.1.2). Furthermore, the Python ecosystem is supported by various packages that extend and enhance the language.

Interactive sessions are available through the iPython shell and the Jupyter notebook.

For example, the NumPy module provides functions that enable the manipulation of numeric arrays and matrices. The SciPy module enables functionality usually expected in scientific computing such as optimisation, linear algebra routines, Fourier transformation, etc.

NumPy provides numerical functions to be used with Python.

The support that Python has for hierarchical modularity makes it possible for programmers and developers to build further functionality. A good example is the Pandas package, which extends the NumPy arrays into dataframes for ease of data manipulation and analysis (see Section 2.5). We will be using Pandas in the rest of the book. Similarly, in this book we will extensively use packages such as Matplotlib, Statsmodels and Scikit-learn which implement plotting, statistical models and machine learning algorithms, respectively.

The Pandas package enables data manipulation and analysis.

Other popular packages are Matplotlib, Statsmodels and Scikit-learn.

It is true that Python is an interpreted language and as such code may be often slower than compiled code tailored to a particular machine architecture. In that respect, although the source code in Python is interpreted on the fly, the main advantage is flexibility. This is an important point in the data science workflow as we are interested in the balance between implementation time versus execution: In many cases we are more interested in getting to grips with the raw data rather than in fine-tuning the execution time for a particular machine.

Python is an interpreted language, in other words, code is read and executed line-by-line.

That brings us to another advantage: Since Python code is not compiled, we have portability. Scripts developed in one environment can be executed in any other one with the appropriate installation. Python is fast enough for the vast majority of the computational tasks in a data science workflow: It is important to get the logic right in the first place and, if needed, make the execution faster later.

Python code is flexible and portable.

It is fast enough for the typical data science workflow

For the purposes of this book, I assume that a suitable Python installation is available to you. Furthermore, I will also assume that the installation is for version Python 3.x rather than Python 2.x. I would like to point out that although version 2 is widely used, more and more users are adopting version 3. Please note that there may be some packages that have not been updated to suit the incompatibilities between the two versions. I hope that in the near future these inconsistencies are ironed out and the community eventually moves to version 3.x.

We will be working with version 3 of the Python distribution.

In particular, I find that the Anaconda distribution[3], built by Continuum Analytics, is robust and complete enough for our purposes. Furthermore, they have suitable distributions for Windows, MacOS and Linux. You can obtain an installation package from `http://continuum.io/` and follow the steps provided by the distribution. Please note that any other Python 3.x installation with the appropriate packages is equally suitable in order to follow the discussions in the rest of the book. I will explicitly mention any modules or packages that are required so that the more seasoned user is able to use pip, easy-install, homebrew or any other installation method they are comfortable with.

[3] Continuum Analytics (2014). Anaconda 2.1.0. `https://store.continuum.io/cshop/anaconda/`

2.1.1   *To Shell or not To Shell*

WE HAVE SEEN ABOVE THAT Python is an interpreted language and as such it is possible to interact with the different commands that have to be executed during the course of a session. Similarly, it is also possible to write all the commands first and execute them in a more conventional development workflow. Both approaches have their advantages and disadvantages, and fortunately you do not have to choose one over the other as Python is flexible enough to enable you to do both.

Python shell enables interactivity. You can also develop scripts to be executed without user intervention.

If you are interested in interacting with the code as you are writing it then starting, for instance, an iPython shell will allow you to type a command and immediately send it to the interpreter for execution. You can then take the output and continue your analysis. This way of working lets us see the results of the commands as we are working. It is an excellent way of prototyping code to be implemented in longer projects.

The interactive shell lets you assess the results of each command you send to the interpreter.

Unfortunately, working in this way makes the code somewhat ephemeral unless you save the commands that you are issuing to the interpreter. If you are interested in keeping track of your development and execute a series of commands repeatedly, then you can save those commands in a text file to create a Python script which by convention has a ".py" extension. The execution of these scripts does not require a Python shell and instead can be executed directly from the command line using the following syntax:

Python scripts let you save the series of commands that form a program. They are saved with the .py extension.

```
python myscript.py
```

where we are assuming that the script called `myscript.py` is saved in the local path. The use of Python scripts makes it easier to launch longer jobs that do not need input from a user to be executed.

The command above is launched directly from the terminal; no need for the Python/iPython shell.

In this book we will take the approach of using an interactive shell (code is compatible with Python/iPython shells) so that we are able to use the output given by the interpreter to explain the different steps we are taking. These individual commands can then be saved in a suitable script that can be run later. We will thus present code as follows:

```
> 42 + 24


66
```

In this book we will present code as used from the Python shell for ease of presentation.

Notice that the diple, >, represents the Python shell prompt where the next line of code is expected. Furthermore, if the command sent to the interpreter expects a printed result, the shell will automatically display it. For ease of explanation, in cases where we do not expect a result, or the discussion requires us to break down the code, we will show it in a script style. In other words, no shell prompt will be shown:

```
a = 42 + 24
```

In cases where we do not expect a result we will not show the shell prompt.

Comments in Python are indicated with the use of the hash symbol, #. The interpreter will ignore any commands that follow until the end of the line. In the example below we enter a comment after carrying out a division operation

A comment in Python is entered with the hash symbol, #.

```
> 2/3    # Python 3

0.6666666666666666
```

Python 3

Please note that the behaviour in Python 2 is different, as the operation above will result in an integer divison:

```
> 2/3    # This integer division returns 0

0
```

Python 2

Since we have passed two integers to the interpreter, Python carries out an integer division, returning only the integer part. If we want a real division we can do the following:

```
> 2/3.0

> 2/float(3)
```

You can avoid integer division by casting integers into floats.

In Python 2 we can import the functionality from Python 3.x with the __future__ module as follows:

```
> from __future__ import division
> 2/3

0.6666666666666666
```

Python 2

Alternatively you can use the __future__ module.

2.1.2   *iPython/Jupyter Notebook*

WE HAVE SEEN HOW THE interactive shell in Python allows us to assess the results of the code we are executing. That may be good enough for a number of tasks, but sometimes it may be desirable to present the code in a way that is easier to interact with, not just for the original programmer, but also with other members of a team or general audience.

The iPython/Jupyter notebook is a great way to do this. Not only does it let you run your code in the same way as the interactive shell and add comments to your code, but also enables you to document your code, calculations and processing all within a web-based interface. In this book, we have made a conscious decision not to use the iPython/Jupyter notebook for printing purposes, however I would encourage you to make use of it whenever you can.

> The iPython/Jupyter notebook lets us interact with the code and also add rich media, text and maths.

The iPython/Jupyter notebook supports the inclusion of text, mathematical expression and inline graphics as well as other rich media such as websites, images, video, maths, etc. At its core, a notebook is a JSON document with the extension `.ipynb`, which makes the files very light and highly portable. The web interface is very handy, and if required, the notebook can be exported to a number of formats such as HTML, LaTeX, PDF, Markdown or even raw Python. Furthermore, the Jupyter project aims to exploit the parts of the code that are not specific to Python and extend their use to other programming languages.

> A notebook is a JSON document, keeping with the general portability principle of Python.

2.2 Firsts Slithers with Python

WE HAVE ALREADY HAD THE opportunity to interact with
the iPython shell and have seen some simple operations
such as addition (+) and division (/). We can continue our
exploration of the programming language as an interactive
calculator. As we would expect, the rest of the arithmetic
operations are supported by Python as shown in Table 2.1.
Notice that the exponentiation in Python is represented
with two stars, **. So far we have used integers and floating
point numbers in the examples presented. It is therefore
natural to ask what other types are supported by Python.

Exponentiation is denoted with **
in Python.

Operation	Operator
Addition	+
Subtraction	-
Multiplication	*
Division	/
Exponentiation	**

Table 2.1: Arithmetic operators in
Python.

2.2.1 Basic Types

AN IMPORTANT FACT TO REMEMBER about Python is that it
is a dynamically typed language. In other words, we do not
need to declare variables before we use them and it is not
necessary to specify their type. Furthermore, each and every
variable that we create is automatically a Python object.

Python is a dynamically typed
language: We do not need to
specify the variable type in
advance.

2.2.2   *Numbers*

As we saw in the previous section, Python supports two types of numbers: Integers and floating point numbers. So we can assign the value of an integer to a variable as follows:

Python supports integers and floating point numbers.

```
> Universe = 42
```

Notice that assignation does not require Python to print anything as a response. We can check the type of an object with the command type:

```
> type(Universe)

int
```

The command type lets us see the type of an object.

Python will let us know what type of object we are dealing with; in this case the object Universe is of type integer. Let us see an example for a floating point number:

```
> Universe2 = 42.0
> type(Universe2)

float
```

2.2.3   *Strings*

A string is effectively a sequence of characters. In Python, strings can be defined with the use of either single (' ') or double quotes ('' '') as follows:

```
> string1 = 'String with single quotes'
> string2 = ''String with double quotes''
> type(string1)


str
```

Strings in Python can be defined with single or double quotes.

In the example above we have requested the type of the variable string1, and as expected, Python tells us that indeed it is a string.

We can ask Python to print a string as follows:

```
> print(string2)


String with double quotes
```

In Python 2 the print statement does not require the brackets.

The + operator is overloaded for strings and it results in the concatenation of these objects:

```
> dead, parrot = ''Norwegian'', ''Blue''
> print(dead + ' ' + parrot)


Norwegian Blue
```

Concatenation of strings can be achieved with the + symbol.

In the example above we have also demonstrated the way in which Python is able to deal with simultaneous assignation in a single line. In other words, the string "Norwegian" is assigned to the variable dead and the string "blue" to the variable parrot. Note that mixing operators between strings and numbers is not allowed, and an error will be thrown if they are. Instead you will have to convert a number to a string using the str function.

Python is able to carry out multiple assignation in a single line. This is part of a Pythonic programming style!

Strings are immutable objects in Python and this means that we cannot change individual elements of a string. We shall discuss more about immutable objects in the context of tuples in Section 2.2.6.

Strings in Python are immutable.

2.2.4   Complex Numbers

PYTHON ALSO SUPPORTS COMPLEX NUMBERS, and it denotes the imaginary number $i = \sqrt{-1}$ as j, and so for a number n, nj is interpreted as a complex number.

In Python, the imaginary number i is denoted with the letter j.

Let us see an example: If we want to define the complex number $z = 2 + 3i$ we simply tell Python the following:

```
> z = 2 + 3j
> print('The real part is {0}, \
    the imaginary part is {1}' \
    .format(z.real, z.imag) )


The real part is 2.0, the imaginary part is 3.0
```

Please note that although the numbers used in the example above are integers, Python recasts them as floating point numbers to suit the complex number object. In the piece of code shown above we have also demonstrated the fact that we can use the backslash (\) to break a line for code readability.

The backslash allows us to break a line.

Remember that each and every entity in Python is an object. Each object has a number of possible actions they are able to perform, i.e. methods. In the example above we have called the `real` and `imag` methods associated with a complex number object to obtain the real and imaginary parts respectively. Another use of a method is shown in the example above for a string, in this case the `format` method to tell Python how to format the printing of a string.

The method of an object can be invoked by following the name of object with a dot (.) and the name of the method.

2.2.5   Lists

A LIST IS PRETTY MUCH self-explanatory: It is a sequence of objects, and these objects can be either of the same type or not. We denote a list with square brackets, [ ]. Lists are mutable objects and therefore it is possible to change individual elements in a list:

A list is denoted by square brackets [ ].

```
numbers = [ 1, 3.14, 2.78, 1.61]
expect = [''Spanish'', ''Inquisition'']
mixed = [10, 8.0, 'spam', 0, 'eggs']
```

It is possible to refer to elements of a list using an index that corresponds to their position in the list:

```
> print(numbers[0])
1

> print(numbers[1:3])
[3.14, 2.78]
```

We can refer to elements in a list with an index.

Indexing in Python starts with the number zero and thus the first element of the `numbers` list is referred to as

Indexing in Python starts at zero.

numbers[0]. Also, we can refer to a sub-sequence of the list using the colon notation as start:end, where start refers to the first element we want to include in the sub-sequence and end is the last element we want to consider in the slice.

Remember that Python interprets the slicing operation up to, but not including, the last item in the sequence. In the example above, Python reads from index 1 and up to index 3, but not including 3. That is why only the second and third elements of the numbers list are returned.

Slicing refers to the subsetting of an array-like object such as lists and tuples.

Since lists are mutable objects it is possible for us to change elements in a list:

```
> expect[0] = 'nobody'
> print(expect)


['nobody', 'Inquisition']
```

We are able to change the elements of a list because they are mutable objects.

We can also add elements to a list with the append method:

```
> numbers.append(1.4142)
> print(numbers)


[1, 3.14, 2.78, 1.61, 1.4142]
```

append lets us add elements to a list.

The new element, 1.4142, is added to the numbers list at the end, increasing the length of the list by one element.

Concatenation of lists is easily achieved with the + operator:

```
> print(numbers + expect)

[1, 3.14, 2.78, 1.61, 1.4142, 'nobody',
'Inquisition']
```

The + symbol lets us carry out list concatenation.

Notice that if the two lists are numerical, the result using the + operator is again the concatenation of the list elements, not the sum.

Another useful method of a list is sort, which does exactly what we would expect: It allows us to sort the list's values. This method will also enable us to see the difference between mutable and immutable objects in our discussion about tuples (Section 2.2.6).

The sort method allows us to sort a list *in place*.

Let us define a list to work with:

```
> List1 = [3, 6, 9, 2, 78, 1, 330, 587, 19]
> print(List1)

[3, 6, 9, 2, 78, 1, 330, 587, 19]
```

We can now invoke the sort method as follows:

```
> List1.sort()
> print(List1)

[1, 2, 3, 6, 9, 19, 78, 330, 587]
```

As we can see using sort with a list results in the elements being ordered in ascending order.

There are a couple of things to note here. First, we have called the sort method using the dot (.) notation. When executing the first line in the code above, the interpreter does not return any values, and that is a good sign: It means that the method executed correctly.

In order to see what happened we issue the second command, which lets us print the contents of List1. As shown above, the elements of the list are now ordered.

This takes us to the second point to note. Since lists are mutable, we can change them and in this case the sort method has changed the elements in List1 to be in ascending order. We have sorted the list *in place*. There was no need to create a copy of the list and sort it.

Since lists are mutable, we are able to change their elements. In this case sorting the elements for instance.

Objects in Python also have functions associated with them. Lists are no exception and in this particular case there is a sorted function too. The difference is that a function will create a new object. Let us take a look:

Lists have a sorted function.

```
> List1 = [3, 6, 9, 2, 78, 1, 330, 587, 19]
> print(sorted(List1))

[3, 6, 9, 2, 78, 1, 330, 587, 19]
```

So far so good, nothing has changed, we end up with a sorted list. However, let us take a look at the List1 object one more time:

```
> print(List1)


[3, 6, 9, 2, 78, 1, 330, 587, 19]
```

As you can see, the object was not changed! Instead, what the sorted function has done is create a new object with the contents of List1 in ascending order.

The sorted function creates a new object with the elements of the original list, but in ascending order.

We could have assigned the result of the function to a new variable and thus create an object that can be referred to at a later stage.

Incidentally, if you require the elements in descending order all you have to do is pass the reverse parameter to either the method:

```
> sorted(List1, reverse=True)


[587, 330, 78, 19, 9, 6, 3, 2, 1]
```

or the function:

```
> List1.sort(reverse=True)
> print(List1)


[587, 330, 78, 19, 9, 6, 3, 2, 1]
```

A very useful pythonic way of constructing lists without the need of a full-blown loop is the so-called *list comprehension*. A typical usage is in the creation of lists whose elements

List comprehension is useful when we need to create a list out of operating on elements of another sequence.

are the result of some operations applied to each member of another sequence or iterable. For example, let us create a string with a sentence:

```
> sentence = 'List comprehension is useful'
> print(sentence)


Lists comprehension is useful
```

We can use the string above to create a list of lists with each word in the sentence in capital and lower-case letters, as well as determining the length of the word. And we can do all this in a single line of code:

```
> words = [[word.upper(), word.lower(), \
    len(word)] for word in sentence.split()]


> print(words)

[['LIST', 'list', 4],
 ['COMPREHENSION', 'comprehension', 13],
 ['IS', 'is', 2],
 ['USEFUL', 'useful', 6]]
```

We are using the string methods split(), upper() and lower() to separate the words in the sentence, and convert them to upper- and lower-case.

2.2.6 Tuples

A TUPLE MAY BE SEEN as a list by another name: They are also sequences of objects, and they may be of mixed type too. They are indeed closely related to lists and apart from the fact that they are defined with round brackets, ( ), the main difference is that tuples are immutable.

Tuples are defined with round brackets ( ).

As we have mentioned above, immutable objects cannot be changed. In other words, we cannot add or remove elements and thus, unlike lists, they cannot be modified in place. Let us take a look at some tuples:

Tuples are immutable objects.

```
> numbers_tuple = (1, 3.14, 2.78, 1.61)
> expect_tuple = (''Spanish'', ''Inquisition'')
> mixed_tuple = (10, 8.0, 'spam', 0, 'eggs')
```

As you can see the only change in the definitions above, compared to the lists in Section 2.2.5, is the use of the round brackets. As with lists, the elements of a tuple can be referred to by their index:

```
> mixed_tuple[4]

'eggs'

> mixed_tuple[0:3]

(10, 8.0, 'spam')
```

Tuples can also be sliced with the help of an index.

Let us see what happens when we try to change one of the elements of a tuple:

```
> expect_tuple[0]='nobody'

TypeError: 'tuple' object does not support item
assignment
```

We are not able to change elements of a tuple as they are immutable objects.

This shows that there are manipulations that are not possible to be done with a tuple. What about sorting? Well, the sorted function still works. Let us define a tuple as follows:

```
> Tuple1 = (3, 60, 18, 276, 87, 0, 9, 4500, 67)
> print(Tuple1)


(3, 60, 18, 276, 87, 0, 9, 4500, 67)
```

We can now apply the sorted function to the tuple:

```
> print(sorted(Tuple1))


[0, 3, 9, 18, 60, 67, 87, 276, 4500]
```

The result of the sorted function on a tuple is a list.

Not too bad, right?, but have you noticed something odd? Well, it seems that the result is not a tuple anymore, but a list! We can see that thanks to the square brackets, and we can make sure of this by using the type command:

```
>   type(sorted(Tuple1))


list
```

This is the result of tuples being immutable: The only way to allow for the elements of the tuple to be ordered is by using the mutable nature of a list. Similarly, since the elements of a tuple cannot be changed, there is no point in having a sort method. Let us have a look:

Since tuples are immutable we cannot change their elements in place.

```
> Tuple1.sort()

AttributeError: 'tuple' object has no attribute
'sort'
```

As stated by the error returned by Python, tuples do not have a sort attribute.

2.2.7 *Dictionaries*

WE ARE ALL FAMILIAR WITH the concept of a dictionary: If we are interested in finding the meaning of a new or unknown word, we simply open up a book (or access a webpage) that lists words in a specified order (alphabetically, for instance). This enables us to search for the word we are interested in. A dictionary in Python serves the same purpose and it is composed of *keys* and *values*.

In the analogy with actual dictionaries, keys are equivalent to words and values are the definitions.

A Python dictionary is defined with the use of curly brackets, { }. Furthermore, the key-value pairs are separated by a colon (:) as follows:

We define a dictionary with curly brackets { }.

```
> dictio = {''eggs'':1, ''sausage'':2,\
    ''bacon'':3, ''spam'':4}
> print(dictio)
{'bacon': 3, 'eggs': 1, 'sausage': 2, 'spam': 4}
```

The keywords in a dictionary can be any immutable Python object including numbers, strings and tuples. The value associated with a particular key can be changed by reassigning the new value to the element of the dictionary

The keys can be any immutable object: numbers, strings or tuples for example.

with the relevant entry. For instance, in our example above we can see that the value to the key spam is 4:

```
> print(dictio['spam'])


4
```

We can change the value of this key by simply reassigning any new value. We can for example reassign the value associated to the spam key:

```
> dictio['spam']='Urggh'
> print(dictio['spam'])


'Urggh'
```

The values in a dictionary can be modified.

This can be done repeatedly:

```
> dictio['spam']='Lovely spam'
> print(dictio['spam'])


'Lovely spam'
```

The modification can be done as many times as required.

It is possible to access the keys and values in the form of straight lists with the aid of the keys() and values() methods:

```
> print(dictio.keys())
dict_keys(['spam', 'bacon', 'eggs', 'sausage'])


> print(dictio.values())
dict_values(['Lovely spam', 3, 1, 2])
```

A list of dictionary keys can be obtained with the keys() method. Similarly, values() returns a list of values in a dictionary.

We can also obtain the key-value pairs in the form of a list of tuples with the items() method:

```
> print(dictio.items())

dict_items([('spam', 'Lovely spam'), ('bacon', 3),
('eggs', 1), ('sausage', 2)])
```

Finally, it is possible to get rid of key-value pairs with the use of the del function:

```
> del dictio['bacon']
> print(dictio)

{'sausage': 2, 'eggs': 1, 'spam': 'Lovely spam'}
```

We can remove entries from a dictionary with del.

2.3   Control Flow

NOT ONLY IS IT IMPORTANT to understand the types and objects that are available in any programming language, but also how to control the flow of a programme to be able to follow the logic behind the way in which the programme itself is organised, in other words, the order in which the individual statements are executed.

In particular it is important to mention that in Python the whitespace is a meaningful character as it enables the definition of blocks of code by having the same level of indentation. Let us see some typical structures to control the flow of a programme in Python.

Whitespace is a meaningful character in Python.

2.3.1 if... elif... else

CONDITIONAL BRANCHING ENABLES US TO perform different actions depending on the result of boolean operations. If a condition is met, then we apply an operation, otherwise a different action is performed. In Python we can do this as follows:

```
if condition1:
    block of code executed
    if condition1 is met
elif condition2:
    block of code executed
    if condition2 is met
...
elif conditionN:
    block of code executed
    if conditionN is met
else:
    block of code executed
    if no conditions are met
```

The if... elif... else... lets us test various conditions and create branches for our code.

As you can see, each block of code is indented at the same level. Also, notice that it is possible to nest various conditions with the help of the elif reserved word. The conditions are logical expressions that can test for scalar comparison and thus we can use any of the comparison operators listed in Table 2.2. Let us see an example:

The conditions to test are logical expression that evaluate to True or False.

Operation	Operator
Equal	==
Different	!=
Greater than	>
Less than	<
Greater or equal to	>=
Less or equal to	<=
Object identity	is
Negated object identity	is not

Table 2.2: Comparison operators in Python.

```
> Age = 40
  if Age > 50:
      print('A wise person')
  else:
      print('Such a youngster')


Such a youngster
```

Finally, remember that the conditions are tested one by one in the order they are provided in the code. If a condition is met, the rest of the tests are not executed.

2.3.2   while

A WHILE LOOP IS USED when we need to repeat a block of code until a condition is no longer met. The structure of a while loop in Python is:

```
while logical_test:
    block of code to be executed
    don't forget to update the test variable
```

An important thing to remember is that at the very beginning of the while loop, the logical test must evaluate to True, otherwise the block of code is never executed. Also, in order to avoid infinite loops we need to update the control variable inside the block of code.

The while loop requires a logical test at the beginning of the block.

We can see how this works by counting down from 10:

```
> countdown = 10
  while countdown >= 0:
      print(countdown)
      countdown -= 1


  10, 9, 8, 7, 6, 5, 4, 3, 2, 1, 0
```

Note that countdown -= 1 is a shorthand for countdown = countdown - 1.

2.3.3 for

TYPICALLY, A while LOOP IS used in cases when we do not know in advance how many times the block of code will need to be executed. If we know how many iterations are needed, we can use a for loop. In Python, a for loop iterates over a sequence: a list, tuple or a string for example.

A for loop is useful when we know how many times the code needs to be repeated.

```
for item in sequence:
    block of code to be
    executed
```

This is the same basic structure used in list comprehension.

The example we used for the while loop in Section 2.3.2 can be written with a for loop as follows:

```
> countdown_list = [10, 9, 8, 7, 6, \
  5, 4, 3, 2, 1, 0]
  for x in countdown_list:
      print(x)


10, 9, 8, 7, 6, 5, 4, 3, 2, 1, 0
```

We could simplify the example above by avoiding the explicit definition of the list and instead define a range:

```
> for x in range(10,-1,-1):
      print(x)


10, 9, 8, 7, 6, 5, 4, 3, 2, 1, 0
```

range enables us to define a sequence of numbers as an object. This means that the values are generated as they are needed.

In the example above we used the range(start, end, step) function to generate a sequence of numbers from start to end-1 in steps given by step. In Python 2, similar behaviour is obtained with the xrange function.

2.3.4 try... except

IT IS NOT UNUSUAL TO have syntactically correct blocks of code with statements that in certain cases may cause an error during execution. These errors are not necessarily fatal in the execution of a programme and instead they are anomalous or exceptional cases that require special processing.

Syntactically correct code may cause errors during execution.

Instances such as those described above are called *exceptions* and when they happen, we are interested in catching them and taking appropriate action, for example by generating an error message. This is what is known as *exception handling*. In Python this can be done with the try...except structure:

```
try:
    Block of code that may raise an
    exception
except Exception1:
    Block of code to run if Exception1
    is raised
except Exception2:
    Block of code to run if Exception2
    is raised
...
except:
    Block of code to run if an unlisted
    exception is raised
```

Exception handling in Python can be done with the try... except structure.

In the structure above exception1, exception2,... are standard exceptions that Python knows about and that are detailed in the appropriate documentation; we list some common ones in Table 2.3.

For further information about standard exceptions see https://docs.python.org/2/library/exceptions.html

For example, we can try to calculate the reciprocal of the elements of a list and print each of the values. However, if the sequence contains the number zero, we can try to catch the exception with ZeroDivisionError:

Standard Exception	Meaning
ArithmeticError	Arithmetic error
FloatingPointError	Floating point operation failure
IOError	I/O operation error
IndexError	Sequence subscript out of range
KeyError	Dictionary key not found
TabError	Inconsistent use of tabs/spaces
UnicodeError	Unicode-related error
ZeroDivisionError	Division by zero

Table 2.3: Standard exceptions in Python.

```
> try:

    for x in range(3,-1,-1):

        print(''The reciprocal of {0} is {1}''.\

        format(x, 1.0/x))

  except ZeroDivisionError:

      print(''Divide by zero? \

      You can't do that!!'')


The reciprocal of 3 is 0.333333333333

The reciprocal of 2 is 0.5

The reciprocal of 1 is 1.0

Divide by zero? You can't do that!!
```

We are using the ZeroDivisionError exception to handle this particular exception.

As you can see, instead of getting an error message and the interpreter halting programme execution, the exception is handled nicely by the code after the appropriate except entry.

2.3.5 *Functions*

Now that we have covered some of the elementary control flow structures in Python, we can start combining them into logical blocks to carry out specific tasks. In particular we can construct pieces of code that can be repeated when necessary and whose outcome depends on the input parameters provided. In other words, we are talking about functions.

A function is a good way to write code that can be repeated, and whose outcome typically depends on inputs provided.

A function in Python has the following syntax:

```
def my_function(arg1, arg2=default2,... \
                argn=defaultn):
    ''' Docstring (optional) '''


    instructions to be executed
    when executing the function


    return result # optional
```

The function definition starts with the word def. Remember that code needs to be indented.

Notice that the function definition starts with the reserved word def and the code inside the function is indicated with appropriate indentation.

The input parameters for the function are the dummy variables arg1, arg2,... , argn and as you can see it is possible to define default values for some of these parameters. Parameters with default values must be defined last in the argument list.

The second line in the function definition is called the *documentation string* and its purpose is to describe the actions that are performed by the function. Finally, notice that it is not necessary for a function to return a result.

A documentation string enables us to provide information about what a function does. Make sure you use it!

Let us define a function to calculate the area of a rectangle sides *a* and *b*:

```
def rect_area(a, b=1.0):
    '''Calculate the area of a rectangle'''
    return a*b
```

We are defining a default value for the parameter *b*.

Notice that the parameter *b* has been given the default value of 1. If we were to call this function with only one parameter, the function will know how to handle the calculations and use the default values when needed.

```
> c = rect_area(20, 2)
> print(c)

40
```

We can use the function by calling it in the same any other in-built Python function is.

In the first line of code above, we are calling the `rect_area` function with two parameters, such that we assign the value 20 to *a* and override the default value of *b* with 2. As expected the area calculated is 40. Let us try providing only one single value to the function:

```
> c2 = rect_area(42.4)
> print(c2)


42.4
```

Here we have only passed the value 42.4 to the function. In this case the value is assigned to *a* and the default value of $b = 1$ is used in the calculation.

We can include control flow structures in our programmes to make them more useful and flexible. Let us for instance implement a simple function to calculate the factorial of a number:

```
def factorial(n):
    '''Return the factorial of n'''
    f = 1
    if n<=1:
        return f
    else:
        while n>0:
            f *= n
            n -= 1
        return f
```

A function can use any of the other control flow structures of the language.

*= and -= indicate repeated operations with the left-hand-side value.

When we pass a number smaller or equal than the one the function expects, it returns the value 1, and when the number is greater than 1 the factorial is calculated with a while loop. Let us use the function:

```
> print(factorial(-3))
    1


> print(factorial(5))
    120
```

There may be times when it is more convenient to define a simple function on-the-fly, without having to resort of a full def structure. In these cases we can exploit the use of the so-called lambda functions:

A lambda function in Python is an anonymous function created at runtime.

```
lambda arg1, arg2, ... : statement
```

where, as before, arg1, arg2,... are the input parameters and statement is the code to be executed with the input provided.

For example, if we needed to calculate the cube of a list of numbers we could try the following code:

```
x = [1, 3, 6]
g = lambda n: n**3
```

In this case the object g is a lambda function that can be called as any other function in Python.

So far nothing too strange: We have initialised a list with the numbers 1, 3 and 6, and then defined a lambda function that calculates the cube of the argument n. We can now apply this function, for example:

```
> [g(item) for item in x]

[1, 27, 216]
```

Lambda functions may seem very simple, but it is that simplicity that provides their strength, as it shown above. This can be seen employed for instance in the implementations of PySpark, the Python API for Spark, an open-source cluster computing framework.

Lambda functions are very useful in frameworks such as Spark.

2.3.6 Scripts and Modules

WITH THE FLEXIBILITY PROVIDED BY the possibility of controlling the flow of a set of instructions, and the repeatability offered by constructing our own functions, it becomes imperative to be able to store programmes in a way that enable us to use and reuse code.

In Python we are able to do this by saving the instructions that make up a programme in a plain text file saved with the extension .py. Furthermore, if we use the interactivity provided by the iPython/Jupyter notebook, it is also possible to save our notebooks in a JSON formatted notebook with the extension .ipynb.

Python scripts have the extension .py whereas notebooks have the extension .ipynb.

It is then possible to execute a Python script from the command line by calling Python followed by the name of the script to be executed. For instance, we can create a script defining a main function and a call to it. We can save the function in a script called firstscript.py with the following contents:

```
def main():
  '''Print the square of a list of
  numbers from 0 to n'''
  n = input(''Give me a positive number'')
  x = range(int(n)+1)
  y = [item**2 for item in x]


  print(y)

main()
```

We are defining a main function in this programme and calling simply with the command main().

In this case we are asking the user for a number n with the command input. We then use this number to calculate a sequence given by the square of the numbers from 0 to n and assign it to the variable y. Finally we simply print the list stored in y.

Notice that we have used n+1 for xrange.

Remember that we have saved the script above, but we have not executed it. We can do this by typing the following command in a terminal in the appropriate path:

```
> python firstscript.py

Give me a positive number: 4


[0, 1, 4, 9, 16]
```

In this case we have given the value $n = 4$ as an input.

This is perhaps not the most advanced algorithm to implement, but we can surely see the possibilities. In particular, we can see how we can create scripts to add further functionality to our code and as such the concept of a module becomes natural.

A module is a file or collection of files containing related Python functions and objects to achieve a defined task. These modules enable us to extend the capabilities of the language, and create programmes that enable us to carry out specific tasks. Any user is able to create their own modules and packages and make them available to others. Some of these modules are readily available for us to be used and once appropriate installation is done all we need to do is import them whenever we need to use them.

> A module is a file containing related Python functions to achieve a specific task.

For example, we can use the `math` module to access some common mathematical functions. Let us create for instance a script that implements a function to calculate the area of a circle. In this case we will need the mathematical constant π to carry out the calculations:

> The `math` module contains some common mathematical functions.

```
import math
def area_circ(r):
   return math.pi * r**2


r=3
Area = area_circ(r)
print(''The area of a circle with '' \
''radius {0} is {1}''.format(r, Area))
```

> We can use the value of π with `math.pi`.

Running the programme will result in the following output:

```
> python area_circ.py

The area of a circle with radius 3 is 28.2743338
```

Notice that we need to tell the Python interpreter that
the constant π is part of the math module by using the
syntax math.pi. In the example above we are importing all
the functions of the math module. This can be somewhat
inefficient in cases where only specific functionality is
needed. Instead we could have imported only the value of π
as follows:

In some cases it may be more
efficient to load only the needed
functionality from a module.

```
from math import pi
```

A large number of modules are available from the Python
Standard Library and more information can be found
in https://docs.python.org/2/library/. In the rest of
the book we will deal with a few of these modules and
packages.

2.4   *Computation and Data Manipulation*

WITH THE PROGRAMMING STRUCTURES DISCUSSED so far
we are ready to take up a large number of tasks, not only in
data science, but in more general settings. In our particular
case, as we shall see in the rest of the book, computation
with data and its manipulation can be managed more

Data manipulation and
computation is a very important
step in the data science and
analytics workflow.

effectively and easily with the aid of linear algebra. In this section we will address some basic concepts in both data manipulation and linear algebra with Python.

2.4.1 Matrix Manipulations and Linear Algebra

AS WE HAVE MENTIONED ABOVE, linear algebra enables us to carry out computational tasks with data in a very effective way. It also provides a compact notation to express the type of manipulations we need to do to our data, from pre-processing to presenting results. The use of vectors and matrices is therefore a very important area to cover. Vectors and matrices are arrays of numerical objects with a defined set of operations such as addition, subtraction, multiplication, etc.

Linear algebra provides us with an efficient and compact way to carry out complex calculations.

An $m \times n$ matrix is a rectangular array of numbers having m rows and n columns. In particular when $m = 1$ we have a column vector and when $n = 1$ we have a row vector. In general, a matrix $\mathbf{A}$ can be represented as follows:

$$\mathbf{A} = \begin{pmatrix} a_{1,1} & a_{1,2} & \cdots & a_{1,n} \\ a_{2,1} & a_{2,2} & \cdots & a_{2,n} \\ \vdots & \vdots & \ddots & \vdots \\ a_{m,1} & a_{m,2} & \cdots & a_{m,n} \end{pmatrix}. \tag{2.1}$$

A matrix can be thought of as a collection of row (or column) vectors.

A Python object that may come to mind when thinking about using arrays is the list. For example, we can create two lists as follows:

See Section 2.2.5 for a discussion about lists.

```
a = [1, 2, 3, 4, 5]
b = [20, 30, 40, 50, 60]
```

However, remember that Python considers these objects
as lists and that each type of object has a defined set of
operations. For instance, if we tried to add these two arrays
in the mathematical sense we will find that Python returns
an unexpected answer:

```
> a + b

[1, 2, 3, 4, 5, 20, 30, 40, 50, 60]
```

Using the + symbol with lists
results in concatenation.

Instead of adding the elements of each of the two vectors,
Python concatenated the lists. This works because Python
has overloaded the + symbol, but an error would be
returned if we tried to use subtraction or multiplication.

We covered list concatenation in
Section 2.2.5.

```
> a - b

TypeError: unsupported operand type(s)
for -: 'list' and 'list'
```

Using other arithmetic symbols
with lists results in an error.

It is clear that a list is a good start for the operations we
need to execute, and the use of the programming
capabilities of Python would enable us to build functions to
define mathematical operations on lists to construct arrays.
However, although it may be a very good programming
practice, rather than building our own functions for this
purpose, we can instead exploit the modules that are

Some useful Python modules for
array calculations are NumPy and
SciPy.

available to us within Python such as SciPy which provides an ecosystem for mathematics, science, and engineering, and in particular NumPy, a package that supports N-dimensional arrays.

2.4.2   NumPy Arrays and Matrices

NUMPY EXTENDS THE TYPES SUPPORTED by Python with the definition of arrays to describe a collection of objects of the same type. The dimension of a NumPy array is defined by a tuple of N positive integers called the *shape* of the array. We can think of arrays as an enhancement on lists and as such we can create arrays with the help of lists:

NumPy extends the types in Python by including arrays.

```
import numpy as np


A = np.array([1, 2, 3, 4, 5])

B = np.array([20, 30, 40, 50, 60])

C = A + B
```

We define a NumPy array with np.array, where np is a convenient alias used for the NumPy package.

In the small piece of code above we are importing the NumPy package and using the alias np to refer to the module. With the aid of the array command in NumPy we transform a list into an array object. If we were to print the content of the array C we would obtain the following:

```
> C


array([21, 32, 43, 54, 65])
```

The use of the + symbol with the arrays defined above results in their addition as expected.

Notice that in this case Python has indeed added the arrays element by element as expected. In the example above we could have simply used the list definitions from the previous section and written the following:

```
A = np.array(a)


B = np.array(b)
```

As we have mentioned above, NumPy extends the functionality of lists in Python to be able to carry out vector arithmetics such as:

- Vector addition: +

- Vector subtraction: -

- Element-wise multiplication: *

- Scalar product: dot()

- Cross product: cross()

These are some of the vector operations that are supported by NumPy arrays.

You may have noticed that we have been referring to vector operations, but what about matrices? NumPy supports matrices too.

```
M1 = np.matrix([[2, 3], [-1, 5]])


M2 = np.matrix([[1, 2], [-10, 5.4]])
```

We can define matrices with the help of np.matrix.

In this case we are using the command matrix to define the objects and the result of the multiplication is as expected:

```
> M1 * M2


matrix([[-28. ,   20.2],

        [-51. ,   25. ]])
```

We can multiply NumPy matrices with the usual multiplication symbol.

An alternative for defining NumPy matrices is to use the mat command to recast NumPy arrays.

A widely used operation in linear algebra is the transposition of a matrix. This can be readily accomplished with the use of the transpose command:

```
> M2.transpose()


matrix([[  1. ,  -10. ],

        [  2. ,    5.4]])
```

Finally, with the SciPy package we can use the linalg methods that will enable us to do some typical linear algebra computations such as matrix inversion:

Linear algebra methods are included in linalg inside SciPy.

```
from numpy import array, dot
from scipy import linalg


x = array([[1, 1], [1, 2], [1, 3], [1, 4]])
y = array([[1], [2], [3], [4]])


n = linalg.inv(dot(x.T, x))
k = dot(x.T, y)


coef = dot(n,k)
```

We can invert a matrix with the .inv method.

Matrix multiplication with arrays can be done with the dot() function.

In the code above we have defined a couple of arrays, x and y. We have then calculated $n = (x^T x)^{-1}$ with the help of the .inv command from the linear algebra module. Note that the command .T in the code returns the transpose of a matrix. We then calculated $k = x^T y$, and finally $coef = nk = (x^T x)^{-1} x^T y$.

We shall come back to this calculation in the context of regression in Chapter 4.

```
> print(coef)

[[ -3.55271368e-15]
 [  1.00000000e+00]]
```

We have deliberately called the result $coef$, as we can think of the result of this simple calculation as the coefficients of a linear regression using the arrays x and y. We will come back to this result in Chapter 4.

2.4.3   Indexing and Slicing

AS IT IS THE CASE with lists, it is possible to access the contents of an N-dimensional array by indexing and/or slicing the array. We can do this using the usual notation start:end:step which will extract the appropriate elements starting at start in steps given by step and until end-1.

Arrays and matrices can be indexed and sliced with the usual colon notation for lists and tuples.

```
> a =  np.arange(10)
> print(a[2:6]); print(a[1:9:3])

[2 3 4 5]
[1 4 7]
```

In the example above we are selecting first the elements from 2 and up to but not including 6. We then ask for the elements from 1 through to 8 in steps of 3.

The same notation can be used with arrays of more dimensions. Let us see an example:

The same applies to arrays of more than 1 dimension.

```
> b = np.array([np.arange(4),2*np.arange(4)])
> print(b.shape)


(2,4)
```

With the shape command we can see the dimensions of matrices and arrays.

The array b above is a 2 × 4 array as can be seen from the shape command. We can select all the elements in row zero as follows:

```
> print(b[:1, :])


[[0, 1, 2, 3]]
```

We are using the colon notation to slice the array. :1 refers to the zero-th row, whereas : indicates all columns.

So far so good, but until now the arrays, matrices and vectors that we have been dealing with have been numerical. However, in many situations the data that we have to deal with is not necessarily all numbers of a single type. There is therefore a need to find a way to accommodate the manipulation of disparate data types, including categorical and text data. In cases like that the capabilities of NumPy are restricted, nonetheless Python can still help as we shall see in the next section.

The arrays we have discussed so far have all been numerical. In many cases we need to deal with different data types and Python can still help.

2.5   *Pandas to the Rescue*

YOU MAY BE THINKING THAT we have lost the plot and
that in the style of the Monty Python troupe we are simply
listing animal names as part of a sketch. You would be
wrong, as Pandas is actually a powerful library that enables
Python to work with structured datasets using panel data
or dataframes. Pandas[4] started life as a project by Wes
McKinney in 2008 with the aim of enabling Python to be a
more practical statistical computing environment.

[4] McKinney, W. (2012). *Python for Data Analysis: Data Wrangling with Pandas, NumPy, and IPython.* O'Reilly Media

Pandas is a great addition to the Python stack: It allows us
to manipulate indexed structured data with many variables,
including work with time series, missing values and
multiple datasets. In Pandas, a 1D array is called a *series*,
whereas dataframes are collections of series. The rich
assortment of data types that can be held by a dataframe,
together with the manipulations that it enables, makes
Pandas an indispensable tool for the jackalope data scientist.

Pandas is a powerful library that enables us to carry out complex data manipulation in a very straightforward manner.

In some sense, we can think of a Pandas series as an
extension of a NumPy array, and indeed we can use them to
initialise a series:

```
import numpy as np
import pandas as pd


array1 = np.array([14.1, 15.2, 16.3])


series1 = pd.Series(array1)
```

A typical alias for the Pandas library is pd.

We could also have used a list or a tuple for initialisation. A very useful feature of Pandas is the ability of using indices and column names to refer to data. Let us consider the data shown in Table 2.4 for some animals detailing their number of limbs and dietary habits:

Animal	Limbs	Herbivore
Python	0	No
Iberian Lynx	4	No
Giant Panda	4	Yes
Field Mouse	4	Yes
Octopus	8	No

Table 2.4: Sample tabular data to be loaded into a Pandas dataframe.

We can load this data into Python by creating lists with the appropriate information about the two features describing the animals in the table.

```
features = {'limbs':[0,4,4,4,8],\
    'herbivore':['No','No','Yes','Yes','No']}


animals = ['Python', 'Iberian Lynx',\
    'Giant Panda', 'Field Mouse', 'Octopus']


df = pd.DataFrame(features, index=animals)
```

We can load data into a Pandas dataframe with lists, dictionaries, arrays, tuples, etc.

Note that we have defined the features `limbs` and `herbivore` from Table 2.4 as a dictionary, where the keys will be the names of the columns in our Pandas dataframe, and the values correspond to the entries in the table. Similarly, we are defining a list called `animals` that will be used as an index to identify each of the rows in the table.

We can have a look at the first three entries in the dataframe df with the command head:

```
> df.head(3)

               herbivore  limbs

Python               No   0

Iberian Lynx         No   4

Giant Panda         Yes   4
```

The head method lets us see the first few rows of a dataframe. Similarly, tail will show the last few rows.

As we mentioned above, we can refer to the column data by the name given to the column. For instance, we can retrieve the data about the number of limbs of rows 2 through to 4 using the following command:

```
df['limbs'][2:5]

Giant Panda     4

Field Mouse     4

Octopus         8
```

We can view the contents of a dataframe column by name, and the data can be sliced with the usual colon notation.

Notice that we have referred to the name of the column as a string. Furthermore, we have use slicing to select the data required. Similarly, the information about a single row can be obtained by locating the correct index:

```
df.loc['Python']

herbivore    No

limbs         0
```

The content of a row can be retrieved with the .loc method.

There is a number of very useful commands in Pandas
that facilitate various tasks to understand the contents
in a dataframe. For example, we can get a description of
the various columns. If the data is numeric, the `describe`
method will give us some basic descriptive statistics such as
the count, mean, standard deviation, etc:

```
> df['limbs'].describe()

count    5.000000
mean     4.000000
std      2.828427
min      0.000000
25%      4.000000
50%      4.000000
75%      4.000000
max      8.000000
```

The `describe` method provides
us with descriptive statistics of
numerical data.

Whereas if the data is categorical it provides a count, the
number of unique entries, the top category, etc.

```
> df['herbivore'].describe()

count     5
unique    2
top       No
freq      3
```

We can also obtain useful
information of categorical data
with `describe`.

It is very easy to add new columns to a dataframe. For
example, we can add a class to our data above as follows:

```
df['class']=['reptile','mammal','mammal',\
   'mammal','mollusc']
```

Adding columns to a Pandas dataframe is very easy.

Pandas also allows us, among other things, to create groups and aggregations:

```
> grouped = df['class'].groupby(df['herbivore'])
> grouped.groups

{'No': ['Python', 'Iberian Lynx', 'Octopus'],
  'Yes': ['Giant Panda', 'Field Mouse']}


> grouped.size()


herbivore

No      3
Yes     2
```

Pandas allows us to group data and create aggregations. The method .groups contains the grouped information, .size returns a simple count.

We can also apply aggregation functions. Let us try to calculate the average number of limbs for herbivores and carnivores in our dataset:

```
> from numpy import mean
> limbs = df['limbs'].groupby(df['herbivore'])\
     .aggregate(mean)
> print(limbs)


herbivore

No      4
Yes     4
```

In this case we are applying the mean function from NumPy to calculate the average per group in our dataset.

In the example above we used Python itself to input data into a Pandas dataframe. Although this is possible for a small dataset, in reality you may be interested to ingest data from other sources. Fortunately Pandas has a very robust input/output ecosystem and is able to take data from a myriad of sources. Table 2.5 lists some of them:

We can import data into a Pandas dataframe from a variety of sources.

Source	Command
Flat file	`read_table`
	`read_csv`
	`read_fwf`
Excel file	`read_excel`
	`ExcelFile.parse`
JSON	`read_json`
	`json_normalize`
SQL	`read_sql_table`
	`read_sql_query`
	`read_sql`
HTML	`read_html`

Table 2.5: Some of the input sources available to Pandas.

Pandas is a very versatile and rich tool and we have only touched the surface in this brief discussion. We will be using Pandas extensively in the rest of the book and whenever possible we will provide explanations to aid the discussion. Nonetheless, we urge you to take a deeper look into this great library.

Pandas is a very versatile library and we will continue using it in the rest of the book.

2.6 Plotting and Visualising: Matplotlib

THEY SAY THAT A PICTURE is worth a thousand words and data visualisation makes the case quite emphatically.

There are a number of tools that enable data visualisation in the context of business intelligence such as Tableau and QlikView or Cognos. In Python, there are some really good modules that support very nice visuals such as Seaborn, or interactivity such as Bokeh. For our purposes we will concentrate on the robustness provided by matplotlib and its Matlab-style API called pylab.

Python is able to create plots and graphs. Here we will cover some of matplotlib's functionality.

In a good pythonic style, matplotlib is an object oriented plotting library that can generate a variety of visualisations: From simple plots, histograms, bar charts, scatterplots and more with a few lines of code. If you are familiar with Matlab or Octave, you will find pylab very easy to use. Let us start by importing the modules:

Matplotlib is an object oriented plotting library. PyLab is a Matlab and Octave inspired API for matplotlib.

```
import numpy as np
import matplotlib.pyplot as plt
from pylab import *
```

In an iPython/Jupyter notebook you can use the magic command %pylab inline to load NumPy and matplotlib.

Let us create a simple figure to plot the following functions:

$$y_1 = x^2, \tag{2.2}$$

$$y_2 = x^3. \tag{2.3}$$

With the aid NumPy we can create a vector with entries for x and calculate y_1 and y_2:

```
x = np.linspace(-5, 5, 200)
y1 = x**2
y2 = x**3
```

The command linspace lets us create an equally spaced vector with a specified number of points.

We can create a plot using the plot command as follows:

```
fig, ax = plt.subplots()
ax.plot(x, y1, 'r',\
   label=r''$y_1 = x^2$'', linewidth=2)
ax.plot(x, y2, 'k--',\
   label=r''$y_2 = x^3$'', linewidth=2)
ax.legend(loc=2) # upper left corner
ax.set_xlabel(r'$x$', fontsize=18)
ax.set_ylabel(r'$y$', fontsize=18)
ax.set_title('My Figure')
plt.show()
```

The commands to create a plot are very similar to those in programming languages such as Octave or Matlab.

Remember that matplotlib is an object oriented library and thus we are using objects to create our plots. The commands above are very similar to those used in Matlab and Octave and should you need to take a closer look at the syntax you can consult other resources[5]. The result of the commands above can be seen in Figure 2.1. Finally, it is possible to save the plot to a file with a single command. In this case we can create a PNG file with the following line of code:

[5] Rogel-Salazar, J. (2014). *Essential MATLAB and Octave.* Taylor & Francis

```
fig.savefig('firstplot.png')
```

2.7   *Summary*

IN THIS CHAPTER WE HAVE covered some of the most important aspects of programming with Python. We started by looking at some of the advantages of using Python in the data science and analytics workflow as well as covering

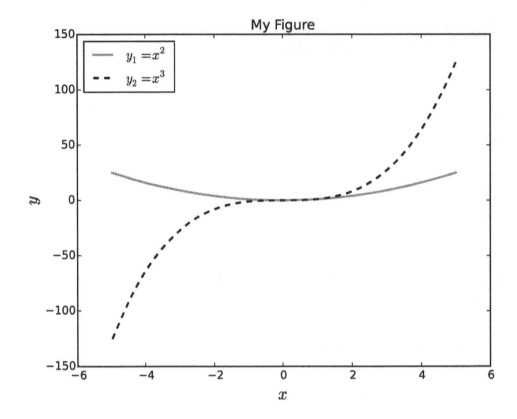

Figure 2.1: A plot generated by matplotlib.

some of the general *pythonista* programming style used by Python programmers.

We have seen how Python can be used as a scripting language as well interactively in a shell such as iPython. We can also use an enriched ecosystem with the help of the iPython/Jupyter notebook. We also covered the different types that are supported by the language: Numbers, strings, complex numbers, lists, tuples, dictionaries. Similarly, we saw how Python deals with mutable and immutable objects.

We can direct the way in which a programme will execute instructions with the help of control flow structures. Furthermore, Python expands its functionality with a number of modules and packages that can be readily imported and used. In this chapter we covered some modules such as NumPy and SciPy. With the help of the powerful Pandas library we can carry out data analysis and manipulation in a very straightforward manner.

Finally, matplotlib is a module that enables us to create plots and visualisations as part of the analysis we perform. These are by no means the only useful modules available to you as a Python programmer, but they are some of the ones we will use throughout of the rest of the book. We will use other modules we have not covered here and we will explicitly mention this in the appropriate sections.

In the next chapter we will cover important concepts from machine learning and pattern recognition that will provide us with the context in which data science and analytics operate. Within that we will present another useful Python library: Scikit-learn.

3

The Machine that Goes "Ping": Machine Learning and Pattern Recognition

WE STARTED OUR DISCUSSIONS ABOUT data science and analytics by stating that the use of data as evidence is nothing new. Furthermore, we saw in Chapter 1 how data science is a portmanteau for a number of overlapping tasks, taking tools from empirical sciences, mathematics, business intelligence, pattern recognition and machine learning. In this chapter we will focus our attention on the latter two in order to provide context to ideas explored so far and frame the algorithms that we will discuss in the following chapters.

3.1 Recognising Patterns

AN INTEGRAL PART OF BEING human is our ability to identify structure and order in the stimuli we receive, aided by information we have acquired previously. From that point of view we are firmly in the realm of *cognitive psychology*, where language usage, memory, creativity and thinking are the main interests[1]. Needless to say, the innate ability to recognise patterns is not unique to humans, and it is general among animals. Do not worry, we are not about to enter into a discussion of perceptual processes or theories of object recognition. However, it may be useful and illustrative to briefly consider the psychological theory of feature analysis.

[1] Eysenck, M. and M. Keane (2000). *Cognitive Psychology: A Student's Handbook.* Psychology Press

In feature analysis, we are said to recognise an object by considering the constituent parts, or features, of the object. We then assemble them together to determine what the object is. For example, we know that a cat is a small fury animal with triangular ears, long whiskers and playful claws. When we see a cat we recognise it for what it is because it satisfies these (admittedly simplified) rules.

If we know what a cat looks like, we can recognise other cats.

This may seem a far-fetched example, were it not for the fact that actual research work on image recognition has been demonstrated using cat faces (and human bodies)[2]. The field of pattern recognition is thus interested in the systematic detection of regularities in a dataset, based on the use of algorithms. We can then use these patterns to take actions such as classifying objects like cats. This does sound familiar, right?

[2] Le, Q. V., R. Monga, M. Devin, G. Corrado, K. Chen, M. Ranzato, J. Dean, and A. Y. Ng (2011). Building high-level features using large scale unsupervised learning. *CoRR abs/1112.6209*

The dataset in question may not necessarily be constrained to the features of a cat, or the parts of a body. As a matter of fact, exploiting patterns has been the bread and butter of sciences such mathematics, physics or chemistry. As a brief example, think of the efforts of Tycho Brahe and Johannes Kepler[3]: One methodically recording the positions of celestial bodies, and the other unraveling the mysteries behind these measurements and summarising them in what we now know as Kepler's Laws.

Recognising patterns is indeed a useful thing to do, and as you may imagine, it is helpful in more than one domain of applications. As a field, pattern recognition started up as part of science and engineering, resulting in a long list of applications to very practical problems.

Nonetheless, engineering was not alone in developing techniques: Other areas such as computer science have also developed capabilities to exploit regularities seen in data. It is therefore an interesting exercise to stop looking for a moment at individual spheres of knowledge and gaze at where they may end up converging. We have talked here about psychology, physics, mathematics, engineering and computer science.

The advances of these domains of knowledge, together with the relentless curiosity of the human mind has posed questions with the aim to understand a variety of areas. In particular, self-reflection has made us turn the attention to ourselves, and has made of the brain a hot area of study. It is a topic that still has many mysteries to reveal.

An important aim in the efforts to understand ourselves is the explanation of how the brain works and how it is able to be the centre of complex activity. This activity manifests

[3] Gilder, J. and A. Gilder (2005). *Heavenly Intrigue: Johannes Kepler, Tycho Brahe, and the Murder Behind One of History's Greatest Scientific Discoveries*. Knopf Doubleday Publishing Group

Pattern recognition is useful in many areas such as science, engineering, computer science, etc.

Self-reflection has made us turn attention to ourselves.

"How does the brain work'?" A very important and difficult question.

itself as creativity, cognition, learning or intelligence, for example.

Is it possible to understand this marvellous organ? And if so, can we replicate its functions? Enter the realm of *Artificial Intelligence*!

3.2   *Artificial Intelligence and Machine Learning*

ARTIFICIAL INTELLIGENCE IS A FIELD that carries a number of connotations: From helpful companion androids through to sentient killer robots, and even the singularity. The key is in the allure of that second word in the compound noun: *Intelligence*. What is intelligence and how can we quantify it? The contempt is due to the first word: *Artificial*. Is it possible to recreate intelligence with the aid of a machine in such a way that the behaviour is similar to that of a person demonstrating intelligence?

The singularity is a hypothetical event where machines are capable of recursive self-improvement leading to an intelligence explosion.

This is not a new aim: The idea of recreating a human-like being has been the inspiration of stories such as that of the Golem, Pinocchio or Frankenstein's monster. It is therefore hardly surprising that when the genius of Alan Turing[4] formulated the concept of a Universal Machine in 1936, the tantalising possibility of an intelligent machine was raised.

[4] Turing, A. M. (1936). On computable numbers, with an application to the Entsheidungsproblem. *Proceedings of the London Mathematical Society* 42(2), 230–265

In the so-called Turing test, Alan Turing considered playing an "imitation game"[5] where a player would have to decide which of two interlocutors is a human being and which is a machine, based only on their written responses to the player's questions.

[5] Turing, A. M. (1950). Computing machinery and intelligence. *Mind* 59, 433–460

If a machine could not be distinguished from the person, then the machine could be said to be "thinking". Indeed, if there is no way of telling what other human beings are thinking except by a process of comparison with one's own thinking, then there is no reason to regard machines any differently.

If we cannot tell what other humans are thinking, there is no reason to regard machines any different in that respect.

As a field, the aim of artificial intelligence is to make machines carry out tasks that are associated with the intellectual processing competence of humans. It is a Herculean labour, and one that not only involves advances in computer science, but also in neuroscience, psychology and even philosophy.

Artificial intelligence aims to make machines carry out tasks associated with intellect.

Among other tasks, pattern recognition, as discussed in the previous section, is an integral part of the various functions that an artificial intelligent agent would have to accomplish. The ability to recognise regularities would enable her to continuously adapt to a variety of changing environmental conditions. This adaptation allows a person to act and react to their surroundings and change their behaviour through learning. The same would be thus expected of the artificial intelligent agent.

Or him...

From that point of view, *machine learning* can be seen as a subfield of artificial intelligence. In that respect, machine learning has a much humbler aim than artificial intelligence: instead of aspiring to the ultimate sentient robot, machine learning is interested in studying the methods that can be used to improve the performance of an intelligent agent over time, based on stimuli from the environment.

Machine learning is a subfield of artificial intelligence focussed on improving the performance of an intelligent agent.

Notice that although this definition uses evocative language, the stimuli may not necessarily be read in real-time or even directly by the intelligent agent, and for that matter the intelligent agent does not have to be necessarily artificial.

Think for example of a business manager who is interested in understanding what marketing materials have the best response in a certain sector of her online customers. The use of data about the browsing habits of her customers could play the role of the stimuli, and this understanding will help improve performance in her business. It is not hard to see why machine learning has become an essential part of data science and analytics.

The use of data generated by businesses can be used to improve their performance.

Machine learning has become ubiquitous in modern life, for example, every time that you check your email and identify a SPAM email in your inbox, you are providing an extra example to an algorithm (quite possibly a naïve Bayes classifier - see Section 6.4.2) that will adapt and learn in order to catch similar SPAM in the future. Similarly, online retailers are able to target products to customers based on items that other similar customers have purchased before with the aid of *collaborative filtering*. We mentioned image recognition earlier on, and other examples include fraud detection, advertisement placement, web search, etc.

The use of machine learning has become ubiquitous in modern life.

See Section 8.4.2 for more information on collaborative filtering.

3.3   Data is Good, but other Things are also Needed

MACHINE LEARNING MAY OFFER US a substantial advantage - and insight - into the problems and questions in our business. In that respect, all we seem to need is an enormous amount of data at our disposal and the rest should follow. Data is indeed an asset in this respect and given current trends, data availability may not be a problem. However, we need to assess whether the data available is indeed relevant to the questions we are seeking to answer.

Data must be treated as an asset.

As you can imagine, it is fairly easy to go down the route that more data is always better. Nonetheless, it may be the case that having access to better relevant data is preferable to having so-called big data. I maintain that any efforts we can spend in improving our data are worth investigating and investing in. After all, the patterns we are trying to exploit can only be as good as the data we employ.

Having relevant data is preferable to having so-called big-data, particularly at early stages.

With that in mind, it is often the case that having a wide variety of data may be more important than having a lot of it. By the same token, being able to apply a variety of clever algorithms may prove to be much more fruitful than simply having rows and rows of raw data. What is more, if the algorithms employed are scalable, adding more data may be a straightforward task.

Using a variety of algorithms is preferable to having lots of data that is not being used.

We mentioned above the need to have relevant data, and the challenge there is to identify when we do indeed have it and when we do not. If we happen to be well-versed in the business domain where the data is being generated, we may have a good chance to decide if it is relevant or not. However, in cases where we do not have such experience, we should be able to face this challenge by enlisting the help of people with experience in the subject matter area. Having

Make sure you have access to relevant subject matter expertise too. It may prove as valuable as the data itself.

that expertise at our disposal can be as valuable as the data itself!

It is therefore recommended to have discussions and reviews with subject matter experts from an early stage in the process. This is particularly true in cases where the data science team may not be fortunate enough to have such expertise.

Additionally, if we are indeed interested in gaining insight from the data, it is also important to discuss the results of the modelling stages with subject matter experts and decision makers. These discussions need to be organised with the understanding that not all of the people involved may be able to follow intricate and lengthy explanations about the finer points of a particular machine learning algorithm.

Communicating your process and results, from the early stages is an important component to becoming a successful jackalope data scientist.

It is therefore important to be able to communicate effectively about the main issues in the process in an inclusive manner. It is only then that the actual effectiveness of the data science process is realised.

3.4   *Learning, Predicting and Classifying*

THE IMPLEMENTATION OF MACHINE LEARNING algorithms involves the analysis of data that could be employed in the improvement (learning) of the agent (model) and subsequently using the results to make predictions about quantities of interest or making decisions in the face of uncertainty.

It is important to bear in mind that machine learning is interested in the regularities or patterns of the data in order to provide predictive and/or classifying power. This is not necessarily the same as causality. We would need a more thorough examination to claim causes and effects given the data we observe.

Machine learning is interested in regularities and patterns in data.

Machine learning tasks are traditionally divided into two camps: Predictive or supervised learning and descriptive or unsupervised learning. Let us start with **supervised learning**: A good example of this type of task is that of a traditional teacher-pupil situation where the teacher presents the pupil with a number of known examples to learn from.

We talk about two types of tasks: supervised and unsupervised ones.

Let us return to the classification of cat faces: A teacher that knows what a cat looks like will present the pupil with several training images of cats and other animals, and the pupil is expected to use the *features* or *attributes* of the images presented to learn what a cat looks like. The teacher will have provided a *label* to each of the images as being of cats or not. In the testing part, the teacher will present images of various kinds of animals, and the pupil is expected to classify which ones show a friendly feline face.

Supervised learning makes use of labelled data.

In machine learning parlance we talk about supervised learning when we are interested in learning a mapping from the input to output with the help of labelled sets of input-output pairs. Supervised learning lets us make predictions based on the data that we see and thus apply **generalisations**.

Supervised learning lets us make predictions.

Each input has a number of features that can be represented in terms of an N-dimensional vector that will help in the task of learning the label of each of the training examples. Think of a supervised learning task as providing an annotated map to a mountaineer that is signing up to our Kilimanjaro expedition and asking her to identify similar landscape features to those marked on the map as she walks along.

Or, where there two mountaineers? Oh well...

The other type of machine learning task is **unsupervised learning**. In this case, following our example of the teacher-pupil situation, the teacher takes a Montessori-style approach and lets the pupil develop, on her own, a rule about what a cat (or any other animal of the pupil's preference) looks like, without providing any hints or labels to the learner.

In this case, from a machine learning point of view, there are no input-output pairs. Instead, we only have the unlabelled inputs and their associated N-dimensional feature vectors, without being told the kind of pattern that we must look for. In that respect, an unsupervised learning task is less well-defined than a supervised one.

In contrast, unsupervised learning does not make use of labelled data

That does not mean that it is less useful, on the contrary, we can use unsupervised learning to gain a better understanding of the data we have acquired and it can provide us with a description or classification of the dataset as well as discovering interesting patterns in the data. In other words unsupervised learning lets us represent our data better by extracting **structure** from it.

Unsupervised learning can help us understand the structure of our data, providing us with ways to describe or classify data points

Under unsupervised learning, in the case of our Kilimanjaro mountaineers, we would ask them to go on their journey without an annotated map, and instead identify interesting areas in the landscape they are able to see from the summit. One important thing to note is that an unsupervised learning task may enable us to assign labels to those inputs and thus open the door to the use of predictive or supervised learning.

> Unsupervised learning may provide us with labels to be used in a supervised learning task.

We have touched upon data being labelled or not, and that has given us some clues as to the sort of problems we can tackle with each of them. Let us now turn our attention to the features and labels (if they exist). In Section 3.1 we gave an example of some of the features that would enable us to recognise a cat. Some of those features can be quantified, for example we mentioned that it must be a small furry animal: How small? We can associate a number to this measurement and then we would be talking about a **numerical** or **continuous** variable. Continuous variables are typically associated with measurement units and we can represent them with real numbers.

> Numerical features are associated with measurement units and represented by numbers.

We may also have attributes that cannot be represented by a number, and instead they provide a description of the type of attribute we are referring to. In the example of the cat we mentioned triangular ears, instead of round or floppy ones. Other examples include colour (black cat, white cat), gender (male/female), etc. We refer to these attributes as **categorical** or **nominal** variables, and are typically related to a class or category.

> Categorical features provide a description of the attributes of the data.

This categorisation among the type of features and labels in a dataset may seem superfluous. However, a further bit of scrutiny will let us see that this innocuous grouping enables us to identify the type of machine learning algorithms that may be more suitable to the problems we would like to tackle. Take a look at Table 3.1, where we provide some quintessential machine learning algorithms that fall into each category.

This distinction between feature types will make it easier for us to understand the sort of machine learning task we want to employ.

Type of learning	Categorical	Continuous
Supervised	Regression	Classification
Unsupervised	Dimensionality reduction	Clustering

Table 3.1: Machine learning algorithms can be classified by the type of learning and outcome of the algorithm.

3.5   Machine Learning and Data Science

MANY PROBLEMS THAT WE WOULD like to tackle using machine learning tend to have high complexity. We have to bear this in mind when trying to apply algorithms, as it is not very likely to find a perfect practical solution. Nonetheless, if the machine can learn so can we. Machine learning algorithms are suitable to the solution of problems encountered in the data science and analytics workflow where we are interested in deriving valuable insights from data.

Remember that if the machine can learn, so can we.

Let us take for instance the case of a supervised learning task where our ultimate aim is to find a function $h(\mathbf{x})$, called the hypothesis. This function enables us to predict values for the problem at hand based on the given input data $\mathbf{x}$. In a practical case, the inputs in the feature vector $\mathbf{x}$ are varied and we would have to decide what the important features to take into account are, and then include them in our model.

We need to decide what are the important features to include in our models.

The optimisation of the predictor $h(\mathbf{x})$ is done using training data points so that for each one of them we have input values $\mathbf{x}_{train}$ that correspond to an output y which is known in advance. Learning, in this sense, is thus the effective use of data in the task of training a model in order to accomplish the job it was set to complete.

In supervised learning we know the output y in advance.

From this point of view, we can relate the tasks that involve that training to the data science workflow steps we listed in Section 1.4: Once we have identified the task at hand, we need to acquire relevant data, extract pertinent features and build our model. In addition to those steps, we also have to consider three important parts that will enable us to decide what sort of machine learning algorithm to choose for our problem.

The steps to follow are all framed by the data science workflow discussed in Section 1.4

With each prediction that we make, we can find the difference between that prediction and the true output value. We do this in order to assess how well out predictor is performing. An important part of the process is to be able to obtain a model that is able to perform well in a general setting, rather than memorising the intricacies of the data provided.

Generalisation is an important outcome of the model.

For example, if we are interested in building an algorithm that is able to recognise cat faces, we would like it to perform well with new, previously unseen cats. If the algorithm is only able to recognise *Bowman*, the Iberian lynx, but not *Mittens*, the kitty, then it is not a great algorithm to be deployed. If, however, the algorithm is able to recognise that a cat is a cat even if it is a drawing, a photograph or an actual real-life cat, then the algorithm is great. We shall come back to this point when we discuss the evaluation of algorithms in Section 3.8.

Model evaluation must be part of the entire process.

At this point, it is pertinent to make clear that there is no such thing as a perfect model, just good enough ones. The improvement in learning comes from generalising regular patterns in the training data to be able to say something about unobserved data points. We should therefore be careful not to obtain a model that "memorises" the data, also known as **overfitting**. We can avoid this by employing techniques such as regularisation and cross-validation as we shall see later on in this chapter.

There is no such thing as a perfect model.

We shall discuss more about avoiding this memorisation in Sections 3.7 and 3.12.

3.6   *Feature Selection*

MACHINE LEARNING CAN BE A powerful tool under our jackalope data scientist belt. Not only can it be used together with computer science, mathematics and statistics to help us filter and prepare our data, but also to extract value out of it. It is therefore important to be able to separate the valuable relationships and patterns from any random, confounding ones. In any real application it is inevitable to have a mix of distracting noise together with the signals we want to exploit.

We have to discern between actual patterns and random noise.

Unprocessed data can thus be thought of as the raw material that can be filtered and prepared to obtain the insights desired. However, as it is the case with cooking, the quality of the ingredients is as important as the steps specified in the recipe. With that analogy in mind, we need to be able to think through the available independent variables or features (ingredients) that will be included in the model (recipe).

Like in cooking, the quality of the ingredients is as important as the steps of the recipe.

In some cases using the unprocessed, raw data may be suitable. However, in many cases it is preferable to create new features that synthesise important signals spread out in the raw data. This process is known as **feature selection** where not only should we consider the the features readily available, but also the creation and extraction of new features and even the elimination of some variables too.

Feature selection considers both existing features, and the creation of new ones.

The careful selection of the features to be used in the modelling helps with the understanding of the model outcomes. It also has a large effect in the predictions obtained from the application of machine learning algorithms. A common way to create new features is via mathematical transformations that make the variables suitable for exploitation by a particular algorithm. For instance, many algorithms rely on features having a linear relationship, and finding a transformation that renders nonlinear features to be represented as being linear in a different feature space is definitely worth considering. We will see some example of this in the next chapter and also in Section 9.1.

Mathematical transformations are a typical way to create new features.

It is true that knowing, a priori, the appropriate transformations and aggregations we should make is a hard task in and of itself. In many cases, experience with similar datasets and comparable applications is invaluable. Nonetheless, if you are starting up not all is lost. Fortunately another common way to extract features is to use machine learning itself.

Experience with data transformations is an invaluable asset to be used.

In this case, unsupervised learning may provide a way to find useful clusters (see Section 5.1) in the data that may point us out in the right direction. Similarly, dimensionality reduction (see Section 8.1) can help us to determine combinations of features that explain the variance shown in our dataset. We shall have an opportunity to talk about these types of algorithms later on in the book.

Unsupervised learning can also be useful in the feature selection process.

3.7   Bias, Variance and Regularisation: A Balancing Act

As we have mentioned in the previous section, machine learning algorithms enable us to exploit the regularities in the data. Our task is therefore to generalise those regularities and apply them to new data points that have not been observed. This is called **generalisation**, and we are interested in minimising the so-called *generalisation error*, i.e. a measure of how well our model performs against unseen data.

Generalisation refers to the performance of a model against unseen data.

If we were able to create an algorithm that is able to recall the exact noise in the training data, we would be able to bring our training error down to zero. That sounds great and we would be very happy until we receive a new batch of data to test our model. It is quite likely that the performance of the model is not as good as a zero generalisation error would have us believe. We have ended up with an **overfit** model: We would be able to describe the noise in our data instead of uncovering a relationship, given the variance in our data.

In principle we can bring our training error down to zero. Unfortunately this translates into a larger generalisation error.

The key is to maintain a balance between the propensity of our model to learn the wrong thing, i.e the **bias**, and the sensitivity to small fluctuations in the data, i.e. the **variance**. In the ideal case scenario we are interested in obtaining a model that encapsulates patterns in the training data, and that at the same time generalises well to data not yet observed. As you can imagine, the tension between both tasks means that we cannot do both equally well and a trade-off must be found in order to represent the training data well (high variance) without risking overfitting (high bias).

The key is to maintain a balance between bias and variance.

High-bias models typically produce simpler models that do not overfit and in those cases the danger is that of underfitting. Models with low-bias are typically more complex and that complexity enables us to represent the training data in a more accurate way. The danger here is that the flexibility provided by higher complexity may end up representing not only a relationship in the data but also the noise. Another way of portraying the bias-variance trade-off is in terms of complexity v simplicity.

Another way to look at this is in terms of complexity versus simplicity of a model.

The tension between bias and variance, simplicity and complexity, or underfitting and overfitting is an area in the data science and analytics process that can be closer to a craft than a fixed rule. The main challenge is that not only is each dataset different, but also there are data points that we have not yet seen at the moment of constructing the model. Instead, we are interested in building a strategy that enables us to tell something about data from the sample used in building the model.

Keeping that balance is more an art than a science.

In order to prevent overfitting it is possible to introduce ways to penalise our models for complexity by adding extra constraints such as smoothness, or requiring bounds in the norm of the vector space we are working on - more on this later on.

This process is known as **regularisation**, and the effects of the penalty introduced can be adjusted with the use of the so-called regularisation hyperparameter, λ.

Regularisation is the process of introducing to our model a penalty for complexity.

Regularisation can then be employed to fine-tune the complexity of the model in question. In a sense it is a way to introduce the Occam's razor principle to our model.

Occam's razor tells us that when we have two competing theories that make the same predictions, the simpler one is preferred.

Some typical penalty methods that are introduced for regularisation are the $L1$ and $L2$ norms that we will discuss in the following section. In Section 3.12 we will touch upon how the hyperparameter λ can be tuned with the use of **cross-validation**.

3.8   Some Useful Measures: Distance and Similarity

ONCE WE HAVE BUILT A set of models based on the training
data we have, it is important to distinguish a good
performing model against a less good one. So, how do we
ascertain that a model is good enough for our purposes?
The answer is that we need to evaluate the models with the
aid of a scoring or objective function.

Various machine learning algorithms have appropriate ways
to let us evaluate how much we can trust what had been
learned and how predictive the model obtained is. The
performance of a model will therefore depend on various
factors such as the distribution of classes, the cost of
misclassification, the size of the dataset, the sampling
methods used to obtain the data, or even the range of values
in the selected features. It is important to note that
evaluation measures are usually specialised to the type of
problem and algorithm used, and the score provided will be
meaningful to the problem domain. For instance, in the case
of classification problems, the classification accuracy may
provide a more meaningful score than other measures.

In general model evaluation can be posed as a constrained
optimisation problem given an objective function. The aim
can then be presented as the problem of finding a set of
parameters that minimises that objective function. This is a
very useful way to tackle the problem as the evaluation
measure can be included as part of the objective function
itself. For example, consider the case where we are

> Remember that we are working with the principle that models are good enough.

> Evaluation measures are usually specialised to the type of algorithm used.

> Model evaluation can be posed as a constrained optimisation problem.

interested in finding the best line of fit given a number of data points: A perfect fit would be found in the case where the data points align flawlessly in a straight line. As you can imagine, that is very rarely the case.

We will discuss regression in Chapter 4.

Instead of expecting the unexpected, we can evaluate how well a line fits the data when we take into account the difference between the location of a point and its corresponding prediction as obtained from the model. If we minimise that distance then we can evaluate and compare various calculated predictions. This particular evaluation measure used in regression analysis is known as the sum of squared residuals (SSR) and we will discuss it in more detail in Chapter 4.

In regression, the minimisation of the sum of squares error is a typical evaluation measure.

As we can see, the concept of distance arises naturally as a way to express the evaluation problem, and indeed a number of conventional evaluation procedures rely on measures of distance. Consider the points A and B in a two dimensional space shown in Figure 3.1. Point A has coordinates $\mathbf{p}(p_1, p_2)$ and point B has coordinates $\mathbf{q}(q_1, q_2)$. We are interested in calculating the distance between these two points. This can be achieved in different ways and we are familiar with some of these, such as the Euclidean and the Manhattan distances.

Various evaluation measures rely on measures of distance.

- **Euclidean distance**: This corresponds to the ordinary distance calculated using the straight line that joins points A and B; in two dimensions it corresponds to the distance given by the Pythagorean theorem. Given the coordinates of each of the two points in question we can

Remember the Pythagoras theorem?

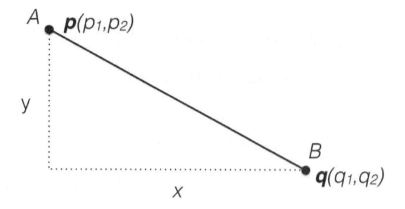

Figure 3.1: Measuring the distance between points A and B.

obtain the distance between A and B as:

$$d_E = \sqrt{(q_1 - p_1)^2 + (q_2 - p_2)^2} = \sqrt{x^2 + y^2}, \qquad (3.1)$$

where the distances x and y are shown in Figure 3.1. It is possible to extend this definition to n dimensions:

$$\begin{aligned} d_E &= \sqrt{(q_1 - p_1)^2 + (q_2 - p_2)^2 + \cdots + (q_n - p_n)^2,} \\ &= \sqrt{\sum_{i=1}^{n} x_i^2}, \qquad (3.2) \end{aligned}$$

This is the well known Euclidean distance.

where x_i is the distance along the i-th dimension. The Euclidean distance is also known as $L2$-norm.

We also call it the $L2$-norm.

- **Manhattan distance**: It is easy to see why this distance measure gets this name if we think of the distance that a yellow cab would cover while travelling along the streets in Manhattan. Apart from Broadway, the cab cannot move diagonally in the street-avenue grid. Instead, it can only move North-South and East-West. In the case of points A and B in Figure 3.1, the Manhattan distance is

To measure the Manhattan distance think of a yellow New York taxi cab and its journey through the island.

given by

$$d_M = |(q_1 - p_1) + (q_2 - p_2)| = |x + y|. \qquad (3.3)$$

For an n-dimensional space we can extend the above definition as:

$$
\begin{aligned}
d_M &= |(q_1 - p_1) + (q_2 - p_2) + \cdots + (q_n - p_n)|, \\
&= \left| \sum_{i=1}^{n} x_i \right|, \qquad (3.4)
\end{aligned}
$$

The Manhattan distance is also known as $L1$-norm.

The Manhattan distance is also known as the L1-norm.

From a geometrical point of view the idea of measuring the distance between two points makes intuitive sense. Furthermore, if the distance is zero we can argue that the two points are effectively the same one, or at the very least *similar* to one another. This idea of similarity is therefore another useful tool in the development of evaluation measures, particularly in the case where features are not inherently amenable to being placed in a geometric space.

Another important concept, related to that of distance, is similarity.

Given two points A and B, the similarity measure d must satisfy a certain number of general conditions:

1. Must be positive: $d(A, B) \geq 0$

2. If the measure is zero, the points A and B are equal and vice versa: $d(A, B) = 0 \longleftrightarrow A = B$

3. Must be symmetrical: $d(A, B) = d(B, A)$

4. Must satisfy the triangle inequality: $d(A, B) + d(B, C) \geq d(A, C)$

The two distance measures we discussed above can be used to gauge similarity, however there are a number of other useful ways to do this, for example the cosine and Jaccard similarities:

Although distance is useful, similarity can be measured in other ways too.

- **Cosine similarity**: This similarity measure is commonly used in text mining tasks, for example. In these cases the words in the documents that comprise the corpora to be mined correspond to our data features. The features can be arranged into vectors and our task is to determine if any two documents are similar or not. Cosine similarity is based on the calculation of the dot product of the feature vectors. It is effectively a measure of the angle θ between the vectors: If $\theta = 0$, then $\cos\theta$ is 1 and the two vectors are said to be similar. For any other value of θ the cosine similarity will be less than 1. The cosine similarity of vectors $\mathbf{v_1}$ and $\mathbf{v_2}$ is given by:

Please note that the cosine similarity is a measure of orientation and not magnitude.

$$s_c(\mathbf{v_1}, \mathbf{v_2}) = \cos(\theta) = \frac{\mathbf{v_1} \cdot \mathbf{v_2}}{|\mathbf{v_1}||\mathbf{v_2}|}, \qquad (3.5)$$

where $|\mathbf{v_i}|$ corresponds to the usual Euclidean norm to measure the magnitude of the vector $\mathbf{v_i}$.

- **Jaccard similarity**: The Jaccard similarity measure provides us with a way to compare unordered collections of objects, i.e. sets. We define the Jaccard similarity in terms of the elements that are common to the sets in question. Consider two sets A and B with cardinalities $|A|$ and $|B|$. The common elements of both sets are given by the intersection $A \cap B$. In order to give us an idea of the relative size of the intersection compared to the sets, we

With Jaccard similarity we can compare unordered collections of objects.

divide the former by the union of the sets. This can be expressed as follows:

$$J(A, B) = \frac{|A \cap B|}{|A \cup B|} = \frac{|A \cap B|}{|A| + |B| - |A \cap B|}. \qquad (3.6)$$

In the case of document similarity for example, two identical documents will have a Jaccard similarity of 1 and those completely dissimilar a value of 0. Intermediate values correspond to various degrees of similarity.

The comparison of documents is a good candidate for the use of Jaccard similarity.

There are other distance and similarity measures that can be used. The choice will depend to a great extend on the type of problem to tackle as well as the algorithms and techniques to be used to solve the problem. In the following chapters we will address specific algorithms and evaluation measures that are appropriate to each of them.

There are other ways to measure distance and similarity, these are some of the most useful/common ones.

3.9   *Beware the Curse of Dimensionality*

WE HAVE BEEN REFERRING TO data features as an integral part of the ingredients we will use with our machine learning algorithms. Once we identified the features to be included in our model we can consider them as the different dimensions along which our data instances can be placed: For a single feature we have a one-dimensional space, two features can be represented in two dimensions, etc.

The features selected to be included in our model can be considered as the different dimensions our data inhabit.

It follows that as we increase the number of features, the number of dimensions that our model must include is

increased too. Not only that, but we will also increase
the amount of information required to describe the data
instances, and therefore the model.

As the number of dimensions increases, we need to consider
the fact that more data instances are required, particularly if
we are to avoid overfitting. The realisation that the number
of data points required to sample a space grows
exponentially with the dimensionality of the space is usually
called the **curse of dimensionality**[6], a term used by Richard
Bellman in the context of dynamic programming, and a
great way to describe this issue!

> As the number of dimensions increases, we need to use more data points to avoid overfitting.

> [6] Bellman, R. (1961). *Adaptive Control Processes: A Guided Tour*. Rand Corporation. Research studies. Princeton U.P

The curse of dimensionality becomes more apparent in
instances where we have datasets with a number of features
much larger than the number of data points. We can see
why this is the case when we consider the calculation of
the distance between data points in spaces with increasing
dimensionality.

> The problem becomes more apparent when we consider the distance between data points in higher dimensional spaces.

Let us consider, without loss of generality, that we have a set
of $M = 10$ data instances belonging to three different classes.
We are interested in finding the closest neighbour to each
of the data points and, in this case, assess if they belong to
the same class or not. This is a very simple classification
task. We can simplify the discussion by considering the use
of a unit length measurement and counting the number
of data points that fall within it. We are depicting this
situation in Figure 3.2 where we show the ten data instances,
represented with a triangle, an open circle and a plus sign,
in 1, 2 and 3 dimensions.

> Let us count the number of data points that fall within one unit length measurement.

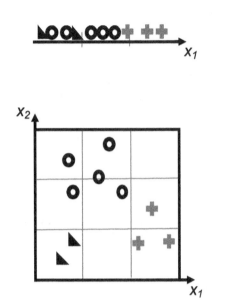

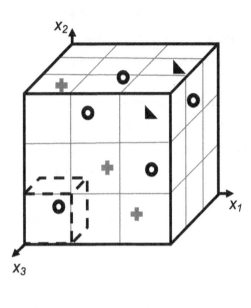

Figure 3.2: The curse of dimensionality. Ten data instances placed in spaces of increased dimensionality, from 1 dimension to 3. Sparsity increases with the number of dimensions.

In a one-dimensional space, the unit measurement is given by a line. In the example shown in Figure 3.2 we can see a space with three unit intervals, so the sample density is $\frac{10}{3}$. In other words we have about 3.333... data points per interval and thus finding a near neighbour and assessing its class is certainly possible.

In the two-dimensional case, we would have to search an area of $3 \times 3 = 9$ unit squares and in this case the sample density is $\frac{10}{9}$, or an average of 1.111... data points per square. In this case it is less likely to find a neighbour of a given data instance within the unit square where the data point of interest is located. Finally, in the 3D case we have a feature space of $3 \times 3 \times 3 = 27$ unit cubes and the sample density is $\frac{10}{27}$ or 0.111... on average. In this case our search for a

As the number of dimensions increases, the distance between data points becomes larger and larger. We end up with very sparse spaces.

neighbour within a cubic square becomes more difficult as most of the feature space is effectively empty. We say that the space is *sparse*.

It is easy to see that as we keep adding features (dimensions), our space becomes sparser. It is due to this sparsity that we require a larger and larger number of data instances when dealing with higher-dimensions. For example, if we were interested in carrying out our classification task using one feature with a range of values between 0 and 1, and we wanted our dataset to cover 25% of this range, we would need 25% of the complete population. By adding another feature we would require 50% of the population ($0.5^2 = 0.25$) and with three features we would need 63% of the population ($0.63^3 \simeq 0.25$).

It is due to this sparsity that we require a larger and larger number of data instances.

We may think that simply adding more data is the appropriate solution to dispel the curse of dimensionality. However, as we saw above, it is important to remember that the number of data instances needed grows exponentially with the number of dimensions. In practice we very rarely have access to an infinite amount of data. Furthermore, using too many features actually results in overfitting.

The number of data instances needed grows exponentially with the number of dimensions.

Following our classification example above with three categories (triangle, open circle and plus sign), it is much easier to find various ways to classify the data instances into separate classes when considering higher number of features. This is a great thing to start with, but we must be careful to avoid overfitting, or even getting carried away with false patterns. Furthermore, the sparsity in a higher

Sparsity in higher dimensional space is not homogeneous.

dimensional space is not homogeneous, and it turns out that the space around the origin is much more sparse than in the corners of the hyper-space.

In order to understand this issue let us consider a 2D space to start with. The mean of the feature space is at the centre of a unit square. If we wanted to search the space within one unit distance from the centre, we would be searching in the area given by a circle of unit radius (circumscribed by the square). Any data instances that fall outside the area of this circle turn out to be closer to the edges of the square and become more difficult to classify as their feature values are more distant to those in the centre (the mean). Let us now consider this situation in N dimensions:

Data instances further away from the centre of a unit circle are more difficult to classify. This is aggravated in higher dimensions.

- The volume of a unit hypercube of N dimensions is
  $1^N = 1$

- The volume of a unit hypersphere of N dimensions[7] is:

$$V(N) = \frac{\pi^{N/2}}{\Gamma\left(\frac{N}{2} + 1\right)} r^N, \qquad (3.7)$$

[7] DLMF (2015). NIST Digital Library of Mathematical Functions. http://dlmf.nist.gov/, Release 1.0.10 of 2015-08-07

where the radius $r = 1$ for the unit hypersphere and $\Gamma(\cdot)$ is the gamma function.

In Figure 3.3 we show how the volume of the hypersphere tends to zero as the dimensionality N increases. Nonetheless, the volume of the hypercube remains fixed. This means that in higher-dimensional spaces, most of the data is actually in the corners of the hypercube that defines the feature space, making the classification task more difficult to achieve.

In higher-dimensional spaces, most of the data is in the corners of the hypercube.

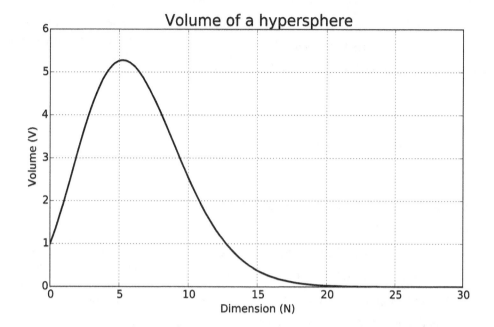

The curse of dimensionality is a very real thing and there is not much that can be done to eliminate it completely. However, it is possible to minimise it, for instance by carefully checking that low-dimensional methods are effective in higher dimensions. Avoiding the curse of dimensionality can be done by increasing the amount of data, but even before going down that route it is worth considering if the features used are indeed a suitable collection.

In that respect, apart from a careful feature selection process, we can also reduce the dimensionality of the problem by transforming the data from a higher-dimensional space into a space with fewer dimensions as it is the case with

Figure 3.3: Volume of a hypersphere as a function of the dimensionality N. As the number of dimensions increases, the volume of the hypersphere tends to zero.

Eliminating the curse of dimensionality completely is not an easy task.

Dimensionality reduction aims to transform data from a higher-dimensional space into one with fewer dimensions.

Principal Component Analysis (PCA). We will discuss this technique in Chapter 8. As for avoiding overfitting, in Section 3.12 we will discuss the ideas behind **cross-validation**. But first we need to make a stop to talk about Scikit-learn.

See Section 8.2 for a discussion on Principal Component Analysis (PCA).

3.10   Scikit-Learn is our Friend

THE BROAD AIM OF MACHINE learning, as we have seen, is the development and application of algorithms that can learn from data. This goal can be accomplished in a variety of ways and in recent times, with the advancement of computer power, together with the availability of useful software tools this is a goal that can be within the reach of many of us.

The advancement of computer power and the availability of useful software tools put machine learning within our reach.

Among the many programming languages and tools that are at hand, in this book we have chosen to use Python. Within the libraries and packages at our disposal we concentrate mainly on Scikit-learn as it contains a wide-range of machine learning algorithms. Scikit-learn builds upon libraries that we have already seen in Chapter 2 such as NumPy, SciPy and matplotlib. Scikit-learn is able to interact with Pandas dataframes and other objects in Python. It is worth mentioning that the focus of Scikit-learn is the modelling part of the data science workflow, rather than the manipulation of data.

Scikit-learn contains a wide-range of machine learning algorithms and builds upon libraries that we have discussed.

Scikit-learn enables us to run popular models and techniques such as:

D,....

- Regression
- Clustering
- Feature selection
- Dimensionality reduction
- Classification
- Cross-validation
- etc.

Some popular models and techniques available in Scikit-learn.

In the following chapters we shall have opportunity to explore some of the implementations of these models. Scikit-learn also comes packaged with some test datasets that can be used for investigating the usage of the various models.

Scikit-learn comes packaged with some test datasets ready to play with.

Given that we will be using this library extensively, it is worth mentioning the typical data representation expected by the models in Scikit-learn. As explained in Section 2.4, matrices and vectors are a favourable data representation, in particular for mathematical calculations and manipulations. Scikit-learn bears that in mind and it expects data to be represented in the form of two-dimensional numeric matrices with M data instances (rows) and N distinct features (columns).

Scikit-learn expects data in a matrix representation.

The matrices can be NumPy or SciPy arrays for instance.

A canonical example in data science and analytics is the *Iris dataset*, and as you can imagine it is included with Scikit-learn. The set was first used by Ronald Fisher[8], and has a total of $M = 150$ samples of three species of iris flowers: Setosa (50), Virginica (50) and Versicolor (50). For each specimen we are provided with $N = 4$ feature measures

[8] Fisher, R. A. (1936). The use of multiple measurements in taxonomic problems. *Annals of Eugenics* 7(2), 179–188

(in centimetres): sepal length, sepal width, petal length and petal width.

We can load this dataset by importing it directly from Scikit-learn as follows:

```
from sklearn.datasets import load_iris

iris = load_iris()
```

With the two lines of code above we have imported the dataset and loaded it into an object named iris. We can now inspect the matrix that contains the feature data. For instance the size of the matrix must be $M = 150$ and $N = 4$, and we can verify this with the shape method as follows:

The Iris dataset is represented by a 150×4 matrix.

```
> iris.data.shape
(150L, 4L)
```

Let us see the first six data instances:

```
> iris.data[0:6,0:4]

array([[ 5.1,   3.5,   1.4,   0.2],
       [ 4.9,   3. ,   1.4,   0.2],
       [ 4.7,   3.2,   1.3,   0.2],
       [ 4.6,   3.1,   1.5,   0.2],
       [ 5. ,   3.6,   1.4,   0.2],
       [ 5.4,   3.9,   1.7,   0.4]])
```

We can use slicing and dicing to see the contents of the dataset.

The dataset also contains the class to which each data instance belongs, i.e. setosa, versicolor or virginica. The information can be obtained by looking at the `target_names` of the `iris` object:

```
> print(iris.target_names)


['setosa' 'versicolor' 'virginica']
```

The class names of the iris flowers.

Remember that Scikit-learn expects data in numeric format, so using strings to represent the classes is not suitable. Instead, each of the three categories has been encoded with numbers corresponding to the position of the names in the list above:

```
> iris.target[0:151:50]


array([0, 1, 2])
```

We will come back to this dataset every so often in the rest of the book.

3.11   *Training and Testing*

THE MODELS WE USE FOR gaining insight into our business or research questions require data. With data as a resource, we need to be mindful of how, when and where it is used. Let us imagine that we are tasked with running a classification model based on the Iris dataset we saw in the previous section. We can consider using all 150 records for this purpose and base our model on the 4 features provided.

We need to be mindful of how, when and where data is used.

We carry out our modelling and the result can be used to classify any new iris flower we encounter based on the 4 measurements (features) used. However, how do we know how well (or how badly) our model performs? We would have to wait until we get new data not seen by the model. This may be an issue as we do not necessarily have a Mark Watney[9] - Martian botanist extraordinaire - or any other (fictional or real) botanist at hand to obtain new iris specimens, whether they are grown on Earth, Mars or elsewhere.

We are interested in the performance against unseen data.

[9] Weir, A. (2014). *The Martian: A Novel.* Crown/Archetype

What is more, we must remember that we build a model because we are interested in using it effectively. This means that we should care about its performance with new unseen data and therefore a way to assess this is with error rates. If we use our entire dataset to train the model, determining our training error becomes impossible. Not to mention the fact that the model will be built to account for the training data only and may thus overfit it, i.e. it will not generalise to new data instances.

We have discussed some causes of overfitting earlier in this chapter.

One way to tackle this problem is to prepare two independent datasets from the original one:

- **Training set**: This is the data that the model will see and it is used to determine the parameters of the model

- **Testing set**: We can think of this as "new" data. The model has not encountered it yet and it will enable us to measure the performance of the model built with the training set

The testing set is sometimes also called the holding set.

In some cases instead of partitioning the data into two sets, it is divided into three. The third component is called the **validation set** and it is used for tuning the model. All three parts need to be representative of the data that will be used with the model. It is important to clarify that the testing data must not be used for training the model, and that the validation set must not be applied for testing.

We can see a schematic representation of this situation in Figure 3.4. Notice that the use of the training set in the modelling provides us with a measure of the training error. In turn, when applying the testing dataset we can get a measure of how well the model performs, i.e. we obtain a measure of the generalisation error. Finally, when using unseen new data we are getting a measure of the so-called out-of-sample error.

From the discussion we had about the curse of dimensionality, we know that the more data instances we have for modelling the better. On the other hand, the more test data we have the more accurate will be our error estimate. A common way to split the data set into training and testing is 80/20% , and typically between one third and one tenth of the data is held out for testing. Other combinations are possible such as 70/30% for example.

In cases when a straightforward splitting leaves us with datasets that may not be representative of the data, we can consider using a stratification method, for instance in situations where a particular class in the data is not represented in the training set. Stratification will aim at

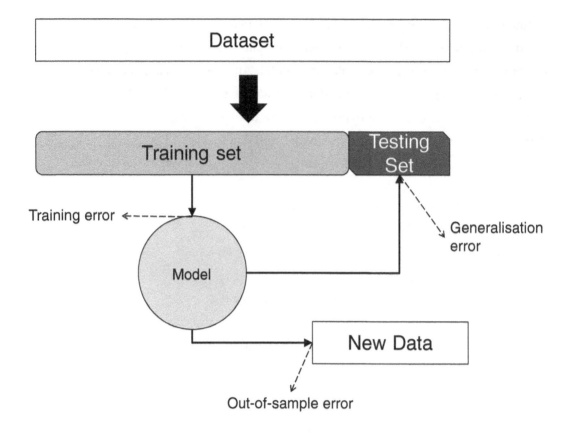

Figure 3.4: A dataset is split into training and testing sets. The training set is used in the modelling phase and the testing set is held for validating the model.

splitting the dataset so that each class is represented in both the testing and training sets.

Splitting the data can be done by partitioning the set into two (or three) sections without altering the order of the data. However, in cases where the dataset has been ordered, this naïve procedure may lead to unbalanced training and testing sets. A better approach is to randomly select the data instances that will form each of the two partitions. It makes sense thus to carry out the randomisation as part of your workflow, regardless of the ordering of the data.

Splitting the dataset into training and testing is better done with random selection.

Sckit-learn is able to assist with the splitting of the dataset into random train and test subsets with the `model_selection` module and its `train_test_split` function following this syntax:

For versions of Scikit-learn earlier than 0.18 this functionality was in the `cross_validation` module

```
model_selection.train_test_split( \
         *arrays, \
         test_size, \
         train_size, \
         random_state)
```

Training and testing datasets can be obtained with the `train_test_split` function.

where `*arrays` are the datasets that will be split, `test_size` accepts a value between 0 and 1 representing the proportion of the dataset to be included in the testing set, `train_size` can be left out, and if so its value is automatically set to be the complement of the test size. Finally, `random_state` initialises the random number generator for sampling.

The train size parameter is optional.

We can see how this can be applied to the case of the Iris dataset. We would like to store the features in arrays called `X_train` and `X_test` and the corresponding targets in `Y_train` and `Y_test`. We would like to hold 20% of the data for testing:

Let us split the Iris dataset with Scikit-learn.

```
from sklearn import model_selection

X_train, X_test,\
Y_train, Y_test =\
   model_selection.train_test_split(\
   iris.data, iris.target,\
   test_size=0.2, random_state=0)
```

We are using the pythonic style of multiple assignation.

We can check the sizes of the newly created sets:

```
> print(X_train.shape, Y_train.shape)

((120, 4) (120,))

> print(X_test.shape, Y_test.shape)

((30, 4) (30,))
```

We can verify the size of the matrices we have after the split.

The modelling task will then be done with the training dataset X_train and Y_train. We can assess how well our model works by measuring the error against the testing dataset X_test and Y_test.

3.12   Cross-Validation

SINCE WE ARE INTERESTED IN making accurate and useful predictions, we need to ensure that any models we create generalise well to unseen data. We have discussed how a training and testing dataset split can help us with this goal. Nonetheless, the parameters that we obtain with the use of a single training dataset may end up reflecting the particular way in which the data split was performed.

In other words, we avoid overfitting.

The solution to this problem is straightforward: We can use statistical sampling to get more accurate measurements. This process is usually referred to as **cross-validation**. Cross-validation improves statistical efficiency by performing repeated splitting of data into training and validation sets, and re-performing model training and evaluation every time. The aim of cross-validation is to use a dataset to validate the model during the training phase.

Cross-validation improves statistical efficiency by performing repeated splitting of data.

Let us see why this helps by considering the following scenario: We have carried out an initial training/testing split. The training set is used for modelling, and we perform the evaluation with the testing set. Imagine now that a different random state had been used to split the data. We would expect the model to see different data points during training. The generalisation error obtained with the second split would be different from the first. We can reduce variability by repeating this scenario over and over again, using different partitions and averaging the validation results over the rounds. Moreover, cross-validation is a great tool when we have a large, but limited amount of data points. Let us see how this can be done with k-fold cross-validation.

Cross-validation can help reduce variability by repeated use of training and testing splits.

3.12.1   k-fold Cross-Validation

A COMMON CROSS-VALIDATION TECHNIQUE IS the k-fold procedure: The original data is divided into k equal sets. From the k subsets, a single partition is kept for validating the model, and $k - 1$ subsets are used for training.

We need to split the dataset into k partitions.

The process is then repeated k times, using one by one each of the k subsets for validation. We will therefore have a total of k trained models. The results of each of the folds can be combined, for instance by averaging, to obtain a

single estimation of the out-of-sample error. We can see
a schematic representation of the k-fold cross-validation
procedure for the case $k = 4$ in Figure 3.5.

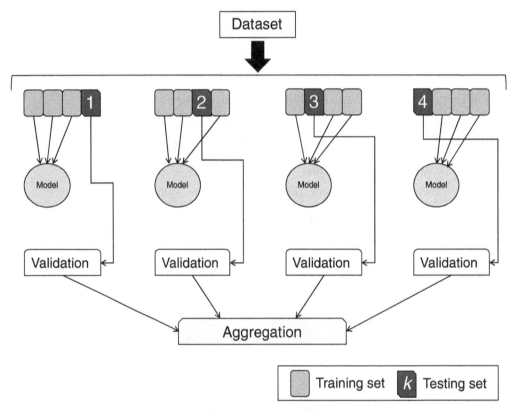

Figure 3.5: For $k = 4$, we split the
original dataset into 4 and use
each of the partitions in turn as
the testing set. The result of each
fold is aggregated (averaged) in
the final stage.

There are other procedures of cross-validation such as
Leave-One-Out, (LOO) where one single data sample is
taken for validation and the rest of the $M - 1$ data points
are used for training. If we were to remove p samples from
the complete set we would be implementing the so-called
Leave-P-Out (LPO) procedure.

Other cross-validation procedures
are also available.

Scikit-learn enables us to carry out cross-validation splits with the aid of functions such as KFold, LeaveOneOut and LeavePOut. The idea behind these functions is to generate k lists of indices that can be used to select the appropriate data points for each fold. For instance, we can create 10 folds for the Iris dataset as follows:

Cross-validation with the k-fold method is implemented with the KFold function.

```
kfindex = cross_validation.KFold(n_splits=10,\
    shuffle=True,\
    random_state=0)


for train_ix, test_ix in kfindex.split(iris.data):
    X_train, X_test =\
    iris.data[train_ix], iris.data[test_ix]
    Y_train, Y_test =\
    iris.target[train_ix], iris.target[test_ix]
```

KFold effectively maintains an index that keeps track of the data instances that go into each of the training and testing sets.

Cross-validation is a useful and straightforward way to get a more accurate estimate of the out-of-sample error, and at the same time a more efficient use of data than a single training/testing split. This is because each record in the dataset is used in both training and validating.

Cross-validation provides a more efficient use of our data.

Cross-validation can also be useful in feature and model selection procedures. For example, it can be used for tuning the regularisation parameter λ introduced in Section 3.7: We split out training data and train a model for a fixed value of λ. We can then test it on the remaining subsets and repeat this procedure while varying λ. Finally, we select the best λ that minimises our measure of error.

Cross-validation can be very useful in tuning the hyperparameter in regularisation.

Despite these advantages we must bear in mind that cross-validation increases the computational work that needs to be done, and if overused, can lead to overfitting. I urge you to use cross-validation given the advantages mentioned above. In a true case of "do as I say, not as I do", there will be examples in the book where I will not perform the cross-validation step as part of the explanations of the various models we will tackle. However, it is worth remembering that cross-validation is an integral part of the modelling part of the data science workflow.

Cross-validation increased the computational work that needs to be done.

3.13   *Summary*

I WOULD LIKE TO FINISH this chapter with a few thoughts that we must always keep in mind during our work in data science and analytics:

- If the machine can learn, so can we!

- Machine learning and data science are not not focussed on causality, but in prediction, insight and knowledge

- All models are wrong: There is no such thing as a perfect model, just good enough ones

- The data science and analytics workflow is a balancing act:

  - Bias v variance

  - Complexity v simplicity

  - Overfitting v regularisation

  - More data v cunning algorithm and resources

  - Accuracy v insight

  - Effort saved v computational cost

  - Jackalopes v unicorns

- Having a lot of data (even big data) is good, and being able to construct models is a great skill. Nonetheless, they are not magic wands

- Beware the curse of dimensionality

- Splitting our data into training and testing not only is good practice but a must

- An important part of the modelling phase in data science is the use of cross validation. Remember that the testing set must never be used for training

4

The Relationship Conundrum: Regression

REGRESSION ANALYSIS IS ONE OF the most widely used tools in statistical analysis. Most of us may have come across it at some point either by employing it or interpreting it. It is a powerful technique due to both its ease of calculation and simplicity of assumptions. However, it is due to these attributes that sometimes regression is misapplied or misinterpreted.

Regression is a well-known and widely used machine learning tool.

In this chapter we will cover the main aspects of regression analysis starting up with a motivation to the problem and covering both linear and polynomial regression techniques. Similarly, we will see how feature selection can be done with the help of appropriate regularisation techniques.

We will cover here some of its most important aspects.

4.1 Relationships between Variables: Regression

CONSIDER A SITUATION WHERE YOU are interested to determine the association between two (or more) pieces of information; say for example the relation of the height of a child compared to that of her parents, or ice cream sales and temperature, or even the body mass of an animal and the mass of its brain. We can collect data for these events and use it for constructing a model that enables us to explore the relationship between the variables in question. Ultimately, our goal is to use our model to predict the outcome of the variable of interest given the values of the other variable(s).

All these are actual, well documented examples of regression usage.

We usually call the quantity of interest the **response** or **dependent** variable and denote it with the variable y. The other quantities are called **predictors**, **explanatory** or **independent** variables and denote them as x. Intuitively, we know that two quantities are correlated if there is a relationship between the two variables, i.e. the value of one tells us something about the value of the other one.

The dependent variable is the quantity we want to predict.

The independent variables are also called regressors.

In a correlation analysis we estimate a value bounded between -1 and 1 and we call it the correlation coefficient. This coefficient tells us the strength of the linear association between the two variables. If the two quantities vary in tandem (if one increases/decreases, the other one does too) the correlation coefficient is positive, whereas it is negative when the two quantities vary out of sync (if one decreases, the other one increases).

The correlation coefficient measures the degree of *linear* relationship among variables.

It is important to remember that the correlation coefficient measures the strength of linear relationship between the variables and therefore a value of zero does not mean that there is no relationship at all. It simply indicates that there is no *linear* relation between the variables in question.

A zero correlation coefficient simply indicates no linear relationship, but other types are available!

Determining the strength of the relationship provides us with some clues towards answering our original question. Although we can tell whether the relationship is strong (± 1) or not (0), the correlation coefficient does not tell us how. A regression analysis does allow us to start seeing how.

Regression analysis lets us explore the relationships among variables.

Before we continue, a word of caution: Just because we measure a correlation between two variables, it does not mean that there is a causal relationship between them. In other words, the fact that people use umbrellas when it rains does not mean that umbrellas cause rain to fall. We are better off avoiding Sir Bedevere's type of reasoning: "If you weigh the same as a duck, then, you're made of wood and must be a witch".

You may have heard the age old aphorism: *"Correlation does not imply causation"*.

Similarly, we must be careful when considering relationships between variables as they may be related to a third, confounding, variable. Take for example the relationship between ice cream sales and temperature we mentioned earlier on: As Summer approaches, the ice cream van is busy selling more ice cones. A similar trend has been noted for the murder rates, as the heat rises, the number of killings do too[1]. In a simplistic analysis one may risk looking at the relationship between ice cream sales and murder, and concluding that one causes the other, without taking into account the weather. Always be on the lookout for confounding variables.

[1] Lehren, A. W. and Baker, A. (2009, Jun 18th). In New York, Number of Killings Rises With Heat. *The New York Times*

Nonetheless, trying to figure out the existence of these relationships and explaining them is by no means something new. As a matter of fact even the name of the technique carries some historical connotations: Sir Francis Galton, a 19th century polymath and first cousin of Charles Darwin, effectively coined the term. Galton was interested in a variety of subjects, from psychology to astronomy as well as statistics. The acceptance of fingerprints as evidence in court was advanced thanks to Galton's studies[2], including estimating the probability that two people have the same fingerprints.

[2] Cole, S. (2004). History of fingerprint pattern recognition. In N. Ratha and R. Bolle (Eds.), *Automatic Fingerprint Recognition Systems*, pp. 1–25. Springer New York

Back to our subject of interest, Galton pioneered the application of statistical methods to many of his scientific interests. For instance, he indeed was interested in the relative size/height of children and their parents[3] (both in animals and plants). Among his observations he noticed that a tall parent is likely to have a child that is taller than average. However, the child is likely to be less tall than the parent. Similarly, a parent that is shorter than average would have children taller than the parent, but still below the average. In other words, the difference in height between parent and offspring is proportional to the parent's deviation from the typical population. He described this by saying that the height of the offspring *regresses* towards a *mediocre* point.

[3] Galton, F. (1886). Regression Towards Mediocrity in Hereditary Stature. *The Journal of the Anthropological Institute of Great Britain and Ireland 15*, 246–263

The height difference in proportional to the parent's deviation from the typical population.

We would call this the mean in modern, politically-correct terms.

Regression towards the mean is a purely statistical phenomenon and can be seen as a fact of life, if you will. The key to the matter is the expectation value of the measured mean: A sprinter that breaks the world record in a race is expected to run to her average time in the next one, or the score in a mid-term exam can be expected to be less bad than the score in the final.

Regression to the mean is an inescapable fact of life.

All in all, regression is thus the mean value of a response variable as a function of one or more explanatory variables. A regression model is an approximate to it. As a first attempt to determining the dependence among the variables, the simplest thing we can do is check if the relationship follows a straight line.

In that sense, a linear regression model assumes (among other things) that the response can be described by a linear function. Even if it is not, we can at least approximate linearly over a range of values or carry out transformations to linearise relationships.

Linear regression assumes:

- Linear relationship.
- Multivariate normality.
- No or little multicollinearity.
- No auto-correlation.
- Homoskedasticity.

The veracity of a linear model may or may not reflect the actual relationship among the variables in question, and we should remind ourselves that there is no such a thing as a perfect model!

The linear regression model has therefore the following form:

$$
\begin{aligned}
\mathbf{y} &= f(\mathbf{x}) + \varepsilon, \\
&= \beta_0 + \beta_1 \mathbf{x} + \varepsilon,
\end{aligned} \tag{4.1}
$$

where β_0 is the intercept of the line, β_1 is the slope of the line, and ε denotes a vector of random deviations or residuals assumed to be independent and identically normally distributed. We refer to β_0 and β_1 as the regression coefficients. In the next section we will extend the model to more than one independent variable and will see how to implement the model using matrix notation.

The intercept is the point where the line crosses the y-axis.

4.2   *Multivariate Linear Regression*

IN THE PREVIOUS DISCUSSION we have only taken into account the relationship between the dependent variable and a single independent one. We can extend the model to include many more variables, for example let us consider N observations on the response y_i with $i = 1, 2, 3, \ldots, N$; and with M regressors $\mathbf{x}_j$ with $j = 1, 2, 3, \ldots, M$. The multivariate linear regression model is written as:

Remember that $\mathbf{x}_i$ are vectors that contain the various data points we will use in our model.

$$y_i = \beta_0 + \sum_{j=1}^{M} \beta_j \mathbf{x}_j + \varepsilon_i. \qquad (4.2)$$

We would like to express the linear regression model in terms of matrices. As such, we can write the independent variable as an $N \times M$ matrix:

$$\mathbf{X} = \begin{pmatrix} 1 & x_{11} & x_{12} & \cdots & x_{1M} \\ 1 & x_{21} & x_{22} & \cdots & x_{2M} \\ \vdots & \vdots & \vdots & \vdots & \vdots \\ 1 & x_{N1} & x_{N2} & \cdots & x_{NM} \end{pmatrix}, \qquad (4.3)$$

The matrix $\mathbf{X}$ provides us with a compact representation of the collection of the different M features for each of our N data points.

whereas the independent variable is a column vector with N elements:

$$\mathbf{Y} = \begin{pmatrix} y_1 \\ y_2 \\ \vdots \\ y_N \end{pmatrix}. \qquad (4.4)$$

Similarly, the vector $\mathbf{Y}$ allows us to collate all the various responses y_i.

Finally, the regression coefficients and the residuals are given by:

$$\beta = \begin{pmatrix} \beta_0 \\ \beta_1 \\ \vdots \\ \beta_M \end{pmatrix}, \qquad (4.5)$$

The intercept β_0 has been included in the regression coefficient vector β.

$$\varepsilon = \begin{pmatrix} \varepsilon_0 \\ \varepsilon_1 \\ \vdots \\ \varepsilon_M \end{pmatrix}. \qquad (4.6)$$

Note that we have included the intercept β_0 in the expression of the regression coefficients. This is why we have a column of ones in the matrix shown in expression (4.3) in the previous page.

In that manner, we end up with the following form for the regression model:

$$\mathbf{Y} = \beta \mathbf{X} + \varepsilon. \qquad (4.7)$$

This is the expression for a multivariate linear regression model.

Our task is therefore to find the regression coefficients in the vector β. The simplicity of Equation (4.7) is provided by the use of matrices. Furthermore, it is also their use that will make all the calculations and manipulations to find the regression coefficients much easier as we shall see in the next section.

4.3   Ordinary Least Squares

WE ARE TASKED WITH FINDING the regression coefficients β
in the multivariate regression model given by Equation (4.7).
Let us recall that we are interested in predicting the value of
the dependent variable given the values of the explanatory
variables. If we were able to craft a perfect linear model,
the actual value of y would match exactly the prediction
$f(x_1, x_2, \ldots, x_M)$. This implies that the residuals ε are zero.

A perfect prediction would make
the residuals equal to zero.

In a more realistic scenario, we would find a pretty good
line of best fit to the data by minimising the error. One way
to implement a suitable objective function for this purpose
is to minimise the sum of squared residuals as follows:

$$
\begin{aligned}
SSR \quad &= \quad \varepsilon^2, && (4.8) \\
&= \quad |\mathbf{Y} - \mathbf{X}\beta|^2, \\
&= \quad (\mathbf{Y} - \mathbf{X}\beta)^T (\mathbf{Y} - \mathbf{X}\beta), \\
SSR \quad &= \quad \mathbf{Y}^T\mathbf{Y} - \beta^T\mathbf{X}^T\mathbf{Y} - \mathbf{Y}^T\mathbf{X}\beta + \beta^T\mathbf{X}^T\mathbf{X}\beta. && (4.9)
\end{aligned}
$$

In linear regression we are thus
interested in minimising the sum
of the square residuals.

Notice that the third term in the last expression is actually a
scalar: $\mathbf{Y}^T\mathbf{X}\beta = (\beta^T\mathbf{X}^T\mathbf{Y})^T$.

Since we require the minimum of the SSR quantity above,
we take its derivative with respect to each of the β_i
parameters. This leads us to the following expression:

$$
\begin{aligned}
\frac{\partial(SSR)}{\partial \beta_i} \quad &= \quad \frac{\partial}{\partial \beta_i}\left(\mathbf{Y}^T\mathbf{Y} - \beta^T\mathbf{X}^T\mathbf{Y} - \mathbf{Y}^T\mathbf{X}\beta + \beta^T\mathbf{X}^T\mathbf{X}\beta\right), \\
\\
&= \quad -\mathbf{X}^T\mathbf{Y} + \left(\mathbf{X}^T\mathbf{X}\right)\beta. && (4.10)
\end{aligned}
$$

We need to take the first derivative
to calculate the minimum.

We can now equate the above expression to zero, leading us to the solution of the matrix equation as:

$$\beta = \left(\mathbf{X}^T\mathbf{X}\right)^{-1}\mathbf{X}^T\mathbf{Y}. \qquad (4.11)$$

This is the solution to the linear model given in Equation (4.7).

We refer to Equation (4.11) as the the normal equation associated with the regression model.

We have already encountered this calculation in Section 2.4.1 where we used Python to demonstrate the use of linear algebra operations such as transposition, inversion and multiplication. Let us go through the calculations one step at a time.

We have actually already implemented this calculation in Section 2.4.1.

4.3.1   The Maths Way

WE CAN USE THE NORMAL equation given in expression (4.11) to solve the linear system given by the linear model in Equation (4.7). Let us see how this is done for the same data used in Section 2.4.1. For the independent variable we have:

We have a single feature and four records.

$$\mathbf{X} = \begin{bmatrix} 1 \\ 2 \\ 3 \\ 4 \end{bmatrix}, \qquad (4.12)$$

and for the dependent variable:

$$\mathbf{Y} = \begin{bmatrix} 1 \\ 2 \\ 3 \\ 4 \end{bmatrix}. \qquad (4.13)$$

We have a very succinct dataset with only four observations and one single feature. In other words, we have an $N = 4$ by $M = 1$ system. Let us express Equation (4.7) as $\beta = \mathbf{M_1}\mathbf{M_2}$. We can now start by calculating $\mathbf{M_1}$ as follows:

In this case we have a 4×1 linear system.

$$\mathbf{M_1} = \left(\mathbf{X}^T\mathbf{X}\right)^{-1}, \tag{4.14}$$

We calculate the first part of the solution as $\mathbf{M_1} = (\mathbf{X}^T\mathbf{X})^{-1}$.

$$= \left(\begin{bmatrix} 1 & 1 & 1 & 1 \\ 1 & 2 & 3 & 4 \end{bmatrix} \begin{bmatrix} 1 & 1 \\ 1 & 2 \\ 1 & 3 \\ 1 & 4 \end{bmatrix}\right)^{-1},$$

$$= \begin{bmatrix} 4 & 10 \\ 10 & 30 \end{bmatrix}^{-1},$$

$$= \begin{bmatrix} 1.5 & -0.5 \\ -0.5 & 0.2 \end{bmatrix}. \tag{4.15}$$

Whereas the second part $\mathbf{M_2}$ is given by:

$$\mathbf{M_2} = \mathbf{X}^T\mathbf{Y}, \tag{4.16}$$

The second part is given by $\mathbf{M_2} = \mathbf{X}^T\mathbf{Y}$.

$$= \begin{bmatrix} 1 & 1 & 1 & 1 \\ 1 & 2 & 3 & 4 \end{bmatrix} \begin{bmatrix} 1 \\ 2 \\ 3 \\ 4 \end{bmatrix}$$

$$= \begin{bmatrix} 10 \\ 30 \end{bmatrix}. \tag{4.17}$$

Finally, the regression coefficients are given by:

$$\beta = M_1 M_2,$$

$$= \begin{bmatrix} 1.5 & -0.5 \\ -0.5 & 0.2 \end{bmatrix} \begin{bmatrix} 10 \\ 30 \end{bmatrix},$$

Finally the multiplication $M_1 M_2$ gives us the coefficients of the linear regression.

$$\begin{bmatrix} \beta_0 \\ \beta_1 \end{bmatrix} = \begin{bmatrix} 0 \\ 1 \end{bmatrix}. \tag{4.18}$$

As we can see from the results above, the intercept of the model is zero and the slope of the line is one. In other words, the model can be expressed by the following equation:

$$Y = x, \tag{4.19}$$

Remember that the model is $Y = \beta_0 + \beta_1 x$.

and therefore the line of best fit is given by a line at $45°$ crossing the origin, as can be seen in Figure 4.1. The grey circles correspond to the data points used for the regression, and the line is given by Equation (4.19).

Not only is it important to be able to carry out the operations that enable us to determine the regression coefficients, but also we must be able to interpret them. In case of the intercept β_0, we can consider this to be the expected mean value of the predicted variable when the independent variable is not present. By the same token, a "unit" increase in the independent variable is associated with a β_1 "unit" increase in the predicted variable.

Not present here means that the dependent variable is zero, $x = 0$.

Please note that in the example used in this section, since all the points considered are perfectly aligned, the line of

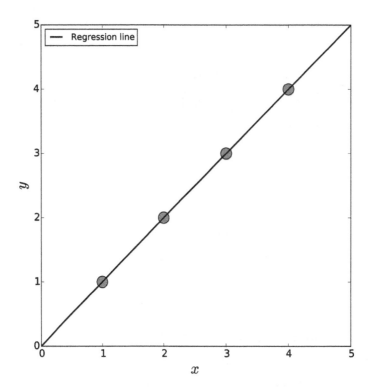

Figure 4.1: The regression
procedure for a very well-behaved
dataset where all data points are
perfectly aligned. The residuals in
this case are all zero.

best fit does indeed pass through every point. However, in
a more realistic situation the presence of noise cannot be
ignored. This is why it is important to get an estimate of the
sum of squared residuals. A schematic representation of this
situation is shown in Figure 4.2, where the distance from
each of the data points to the line of best fit is shown as a
dot-dashed line.

Although the calculations we have covered above are
straightforward, they can be somewhat laborious. This is
particularly true in the case where there are more than one

The calculations shown are
straightforward, but can get
somewhat laborious.

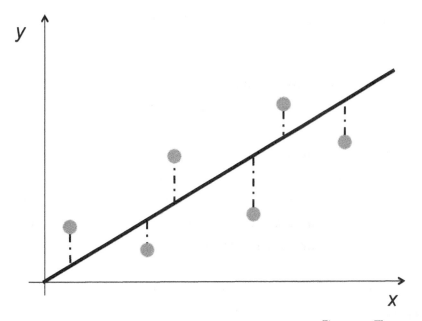

Figure 4.2: The regression procedure for a very well-behaved dataset where all data points are perfectly aligned. The residuals in this case are all zero.

or two independent variables at play. Furthermore, the matrix inversion that must be carried out is not a trivial calculation, and we should be careful as there is no guarantee that a given matrix is invertible. In those cases, we say that the matrix is singular or degenerate and methods to approximate the inverse are needed.

We can implement a function in Python with the normal equation given by expression (4.11). A naïve approach may be well suited for simple problems. In a nutshell, we do not really want to solve the problem "by hand" as done in this section, but perhaps we are better off using Python libraries that are readily available for this task such as those in StatsModels and Scikit-learn.

Instead of doing the computations "by hand", we are better off using a computer.

4.4   Brain and Body: Regression with One Variable

LET US NOW TAKE A look at running a linear regression
with a slightly larger dataset. In this case, we are going
to follow one of the examples that we mentioned at the
beginning of this chapter. The dataset that we will use
looks at relationship of the body mass of an animal and
the mass of its brain[4]. The data is available at `http://`
`dx.doi.org/10.6084/m9.figshare.1565651` as well as at
`http://www.statsci.org/data/general/sleep.html`.

[4] Allison, T. and D. V. Cicchetti
(1976, Nov 12). Sleep in mammals:
ecological and constitutional
correlates. *Science 194*, 732–734

We will assume that the data has been downloaded into a
comma-separated-value (CSV) file with the name
`mammals.csv` and saved in a subfolder called `Data`. We can
use Pandas to manipulate the file. Let us start by looking at
a scatter plot of the data. Before we can do this we need to
upload the necessary modules:

```
%pylab inline
import numpy as np
import matplotlib.pyplot as plt
import pandas as pd
```

We start by importing the
Python libraries that will help
us manipulate the data.

Note that the `%pylab inline` command at the beginning of
the code above imports NumPy and matplotlib enabling
plots to be printed in the Jupyter notebook. We are also
explicitly importing these libraries to make the code a bit
clearer. Finally, we also import Pandas, and assign to it the
alias `pd`.

The `%pylab inline` command
imports NumPy and matplotlib.

Let us load the data into a Pandas dataframe called `mammals` and visualise the data with a scatter plot as shown in Figure 4.3:

```
mammals = pd.read_csv(u'./Data/mammals.csv')
plt.scatter(mammals['body'], mammals['brain'])
```

Using the `read_csv` method from Pandas we can read our CSV file.

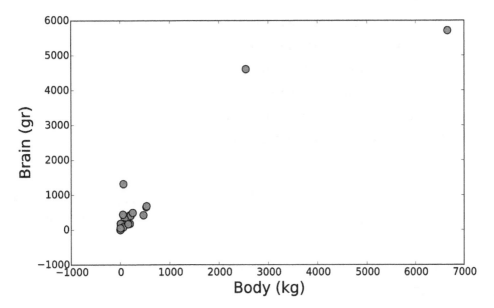

Figure 4.3: A scatter plot of the brain (gr) versus body mass (kg) for various mammals.

Let us take a look at the number of entries for each of the two variables we are considering. First the variable called body:

```
> body_data = mammals['body']
> body_data.shape

(62, )
```

The variable `brain` gives us the following:

```
> brain_data = mammals['brain']
> brain_data.shape


(62, )
```

As we can see there are 62 records in the dataset. We have assigned the values of each of the columns of the dataset to new variables for ease of reference. Here we will make use of the StatsModels module to perform the regression. We show how to use Scikit-learn in Section 4.4.1. First we have to add a column of ones to the body variable so that we can accommodate for the calculation of the intercept, β_0. StatsModels has a handy function for this purpose: `add_constant`

The shape command allows us to see the size of the dataset imported into Pandas.

We make use of StatsModels here. You can see an example of using Scikit-learn in Section 4.4.1.

```
import statsmodels.api as sm
body_data = sm.add_constant(body_data)
```

We make use of the StatsModels library to carry out the regression.

In the code above we are loading the StatsModels package and using the `sm` alias to refer to it. We are then adding a column of ones to the `body_data` array making it a 62×2 array. We are ready to run our regression with the ordinary least squares method (OLS) implemented in StatsModels:

```
regression1 = sm.OLS(brain_data, body_data).fit()
```

In particular we are using the ordinary least squares method implemented in StatsModels.

If you are familiar with R, you are probably familiar with the "formula notation" used to refer to the dependency of a variable on appropriate regressors. So, if you have a

function $y = f(x)$, in R you can denote that dependency
with the use of a tilde, i.e ~. This is an easy way to deal with
denoting the dependency among variables and fortunately
StatsModels has an API that allows us to make use of it in
Python too:

```
import statsmodels.formula.api as smf
regression2 = smf.ols(formula =\
    'brain ~ body',\
    data = mammals).fit()
```

The formula API in StatsModels
allows us to simply the notation
in the linear regression. It uses
formulas similar to those in R.

Please note that the R-style formula does not require us to
add the column of ones to the independent variable. The
regression coefficients obtained with both methods are the
same, and do not forget that the `.fit()` command is needed
in both cases.

Do not forget to use the `fit()`
command to carry out the
regression.

Let us now take a look at the results of fitting the model to
the data provided. StatsModels provides a summary
method that renders a nice-looking table with the
appropriate information. Unfortunately the formatting of
the output is more suitable for showing on a computer
screen rather than in a book page. Nonetheless, you are
encouraged to try the following command in your shell:

```
regression1.summary()
```

StatsModels provides a nice
summary of the regression using
the `summary()` command. The
output is more suitable to be
shown on a computer screen: go
ahead and try it in your Python
implementation.

You can run the same command for the second model we
executed and you will see the same results:

```
print(regression2.summary())
```

Among the information provided you will have the following entries:

```
OLS Regression Results
==========================================
Dep. Variable:                     brain
Model:                               OLS
Method:                    Least Squares
No. Observations:                     62
R-squared:                         0.873
Adj. R-squared:                    0.871
```

Part of the summary provided by OLS gives us information about the goodness of fitness via the R^2 coefficient.

OLS tells us the name of the dependent variable, the model used and the method as well as the number of observations used. It also tells us the value of R^2 also called the coefficient of determination. The values of this number range between 0 and 1, and it tells us how well the data fit the model. A value of 1 is an indication that the regression line obtained fits the data perfectly well, whereas a value of 0 tells us that the linear model is no good. This measure is related to the Pearson correlation coefficient between the dependent and explanatory variables. We could formulate the linear regression model as a maximisation problem for R^2.

The R^2 coefficient is related to the Pearson correlation.

Having said that, there are some drawbacks with only looking at the value of R^2. Namely, that it increases as we add more explanatory variables to the mix. We should therefore be careful when running regression models by adding extra features: An increase in the value of R^2 may not be due to the explanatory power of the input, but to the fact that we added that extra input. That is why OLS also provides information for the adjusted R^2. It is very similar to R^2, but it introduces a penalty as extra variables are included in the model. The adjusted R^2 value increases only in cases where the new input actually improves the model more than would be expected by pure chance.

Although the coefficient of determination provides an indication about how well the model fits the data, it should be used with care.

For the case of our dataset, an $R^2 = 0.873$ is a pretty good outcome as 87% of the total variance in brain mass is explained by the linear regression model based on the body mass. This means that the regression line obtained must be a good one. And as yet we have not even mentioned what this line is and how it can be obtained from OLS.

An alternative is the adjusted R^2 value.

	coef	std err	t	P>|t|	95% Conf. Interval	
Intercept	91.0044	43.553	2.090	0.041	3.886	178.123
body	0.9665	0.048	20.278	0.000	0.871	1.062

Table 4.1: Results from the regression analysis performed on the brain and body dataset.

In Table 4.1 we can see the regression parameters obtained from running the ordinary least squares method on the brain and body dataset. The column named "coef" lists the estimated values of the coefficient listed in the table. Notice that the "const" corresponds to the intercept of the model.

OLS lists the rest of the coefficients using the names of the variables included in the model. The "std err" column corresponds to the basic standard error of the estimate of the coefficient; "t" is the so-called t-statistic and it tells us how statistically significant the coefficient is. The P-value is listed in the "P > |t|" column and it helps us determine the significance of the results considering the null-hypothesis that the coefficient being equal to zero is true. A small P-value (typically ≤ 0.05) indicates strong evidence against the null hypothesis and you should go with the value obtained for the coefficient. Finally "95% Conf. Interval" gives us the lower and upper values of the 95% confidence interval.

OLS lists the names of the variables as they appear in our data.

A small P-value indicates evidence against the null hypothesis and so it is rejected.

The results shown in Table 4.1 indicate that the intercept for the model is $\beta_0 = 91.0044$ and the slope of the line is $\beta_1 = 0.9665$, leaving us with the following model:

$$Brain = 0.9665(Body) + 91.0044, \qquad (4.20)$$

and the P-values obtained indicate a rejection of the null-hypothesis. We can get the parameters using the params method:

```
> regression2.params

Intercept        91.004396
mammals.body      0.966496
```

The regression parameters can be obtained with the params method of the fitted model object.

We can use this equation to predict the mass of a mammal given its body mass and this can easily be done with the predict method in OLS. Let us consider new body mass measurements that will be used to predict the brain mass using the model obtained above. We need to prepare the new data in a way that is compatible with the model. We can therefore create an array of 10 new data inputs as follows:

```
new_body = np.linspace(0,7000,10)
```

For the `predict` method of the model run with the formula API we do not need to add a column of ones to our data and instead we simply indicate that the new data points are going to be treated as a dictionary to replace the independent variable (i.e. `exog` in StatsModels parlance) in the fitted model. In other words, you can type the following:

The `predict` method of the StatsModel formula API does not need the addition of a column of ones.

```
brain_pred=regression2.predict(exog=\
    dict(body=new_body))
print(brain_pred)
```

`exog` refers to the independent variable.

This will generate the following output:

```
array([  91.00439621,    842.72379329,
       1594.44319036,   2346.16258744,
       3097.88198452,   3849.6013816 ,
       4601.32077868,   5353.04017576,
       6104.75957284,   6856.47896992])
```

The numbers shown correspond to the brain mass predictions for the artificial body mass measurements used as input. In Figure 4.4 we can see the regression line given by Equation (4.20) in comparison to the data points in the

We can use the values above to construct the line of best fit given by the model.

set. Please note that if you are not using the formula API,
the input data will require the addition of a column of ones
to obtain the intercept.

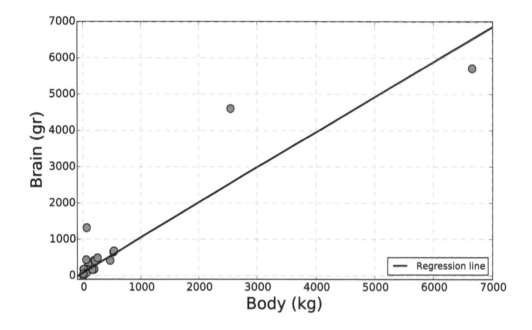

Figure 4.4: A scatter plot and the
regression line calculated for the
brain (gr) versus body mass (kg)
for various mammals.

So far so good, but can we do better than that? For example,
look at the clustering that happens in the region below
$1,000$ kilogram mark for body mass and compare it with the
ones that take place after the $2,000$ or $6,000$ kilogram marks.
Are the latter outliers?, or can we come up with a better
model that encompasses these differences? Let us take a
look at a typical transformation carried out in a variety of
analyses. But before let us use Scikit-learn.

4.4.1 Regression with Scikit-learn

WE HAVE SEEN HOW TO use StatsModels to perform our linear regression. One of the reasons to use this module is the user-friendly output it generates. Nonetheless, this is not the only way available to us to perform this analysis. In particular, Scikit-learn is another very useful module, and one that we use extensively throughout the book. For completness, in this section we will see how to perform linear regression with Scikit-learn.

We can implement a regression model using Scikit-learn.

Let us import the modules to read the data. We will use Pandas to load the data into a dataframe called `mammals`:

```
%pylab inline
import numpy as np
import pandas as pd


mammals = pd.read_csv(u'./Data/mammals.csv')
```

This is exactly the same we did in the previous section.

As we discussed in Section 3.10, we know that Scikit-learn expects data to be represented by two-dimensional numeric matrices with M data instances (rows) and N distinct features (columns). In this case, we have 62 instances and one feature. Let us arrange the data as expected by creating appropriate arrays for the dependent and independent variables:

```
body_data = mammals[['body']]
brain_data = mammals[['brain']]
```

Notice the use of double brackets to get the right shape for the arrays.

We are now ready to create our model. First, we need to create an instance of a linear regression model from `linear_model` in Scikit-learn. This can be done as follows:

```
from sklearn import linear_model
sk_regr = linear_model.LinearRegression()
```

We carry out linear regression modelling with `linear_model` in Scikit-learn.

With that in place we simply use the `fit` method for the model and we are done:

```
sk_regr.fit(body_data, brain_data)
```

We can now check that the intercept and coefficient obtained are the same we calculated with StatsModels. We can also check the value of the R^2 coefficient:

```
> print(sk_regr.coef_)
[[ 0.96649637]]

> print(sk_regr.intercept_)
[ 91.00439621]

> print(sk_regr.score(body_data, brain_data))
0.872662
```

The coefficients and intercept are obtained with the `.coef_` and `.intercept_` methods.

The R^2 coefficient is calculated with the `.score` method.

Finally, the predictions can be calculated with the `predict` method:

```
new_body = np.linspace(0, 7000, 10)
new_body = new_body[:, np.newaxis]
brain_pred = sk_regr.predict(new_body)
```

Note the use of `np.newaxis` to render the Numpy array into the format expected by Scikit-learn.

4.5   *Logarithmic Transformation*

ONE OF THE PRINCIPAL TENETS of the linear regression
model is the idea that the relationship between the variables
at play is linear. In cases when that is not necessarily true,
we can apply manipulations or transformation to the data
that result in having a linear relationship. Once the linear
model is obtained, we can then undo the transformation to
obtain our final model.

Hence the name!

A typical transformation that is often used is applying
a logarithm to either one or both of the predictive and
response variables.

Applying the logarithmic function
is a typical way to transform our
data.

Let us see what happens to the scatter plot of the body
and brain data we have been analysing when we apply the
logarithmic transformation to both of the variables. We will
create a couple of new columns in our Pandas dataframe to
keep track of the transformations performed:

```
from numpy import log
mammals['log_body'] = log(mammals['body'])
mammals['log_brain'] = log(mammals['brain'])
```

Remember that `log` in Python
refers to the base-*e* logarithm.

We can plot the transformed data and as we can see from
Figure 4.5 the data points are aligned in a way that indicates
a linear relationship in the transformed space.

Why has this happened? Well, remember that we are trying
to use models (simple and less simple ones) that enable us
to exploit the patterns in the data. In this case the

The transformation has helped us
convert our problem into a simpler
one.

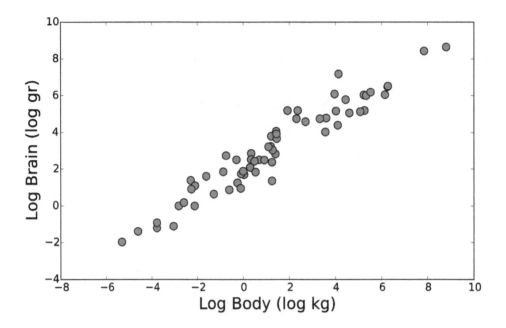

Figure 4.5: A scatter plot in a log-log scale for the brain (gr) versus body mass (kg) for various mammals.

relationship we see in this data may be modelled as a *power law*; e.g. $y = x^b$.

The log-log transformation applied to the data maps this nonlinear relationship to a linear one, effectively transforming the complicated problem into a simpler one:

$$\log(y) = \log(x^b), \tag{4.21}$$
$$\log(y) = b\log(x),$$
$$y' = bx', \tag{4.22}$$

where we are using the notation log for the inverse of the exponential function e. As we can see, we have transformed

In engineering and other disciplines they use the notation ln for this function.

our power law model, a nonlinear model in the regressor, into a form that looks linear as shown in Equation (4.22).

Now that we have this information at hand, we can train a new model using the transformed features attached to the `mammals` dataframe:

```
log_lm=smf.ols(formula = 'log_brain ~ log_body',\
    data = mammals).fit()
```

The only difference here is that we are using the transformed variables in the OLS function.

If we print the `log_lm.summary()` we will see some of the following information:

```
OLS Regression Results

========================================

Dep. Variable:              log_brain
Model:                            OLS
Method:                 Least Squares
No. Observations:                  62
R-squared:                      0.921
Adj. R-squared:                 0.919
```

However, the results returned by OLS are remarkably different.

The logarithmic transformation has increased the value of R^2 from 0.873 to 0.921. We can see the value of the sum of squared residuals with the following command:

```
> log_lm.ssr


28.9227104215
```

We can obtain the sum of the square residuals with the `ssr` method.

Let us now take a look at the statistics for the model as well as the all important coefficients. As we can see from Table

	coef	std err	t	P>|t|	95% Conf. Interval	
Intercept	2.1348	0.096	27.227	0.000	1.943	2.327
log_body	0.7517	0.028	26.409	0.000	0.695	0.809

Table 4.2: Results from the regression analysis performed on the brain and body dataset using a log-log transformation.

4.2, the new model has an intercept of $\beta_0 = 2.1348$ and a slope of $\beta_1 = 0.7517$. The regression line for this model can be seen in Figure 4.6.

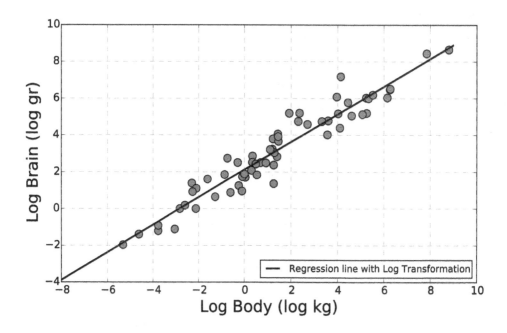

Figure 4.6: A log-log scale figure with a scatter plot and the regression line calculated for the brain (gr) versus body mass (kg) for various mammals.

Remember that the coefficients obtained above are for the transformed data and if we wanted to relate this to the original features we would need to undo the transformation. In this case we have a model given by

$$Brain = A(Body)^{0.7517} \qquad (4.23)$$

where $A = e^{2.1348}$.

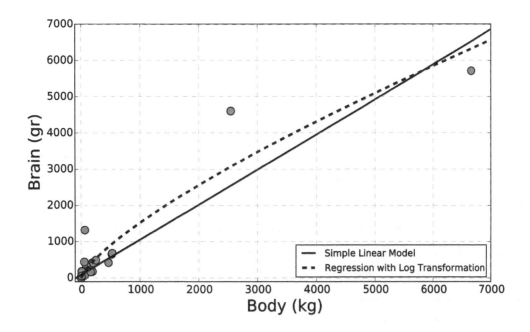

Figure 4.7: A comparison of the simple linear regression model and the model with logarithmic transformation for the brain (gr) versus body mass (kg) for various mammals.

In Figure 4.7 we can see the original scatter plot and a comparison of the two models. It is easy to see how the logarithmic transformation allows for greater flexibility, capturing those data points which are a struggle for the simple linear model. This comparison demonstrates that it is possible to build a variety of models to explain the behaviour we observe in the data. In Section 4.7 we will

The logarithmic transformation enabled us to capture those data points which are a struggle for the linear model.

see how we can fit a polynomial regression to the same dataset. Nonetheless, please remember that carrying our appropriate splitting of training and testing sets, together with cross-validation is an unrivalled way to decide which model, among those tried, is the most suitable to use with unseen data.

4.6   Making the Task Easier: Standardisation and Scaling

GIVEN THAT THE MAIN UNDERLYING concept behind linear regression is the assumption of a linear relationship, transformations such as the one covered in the previous section make the task easier for both the learning algorithm and for us. As you can imagine, there may be many more tricks up our jackalope sleeves to transform and pre-process the data in order to facilitate our modelling. In this section we are going to present a couple of the most widely used techniques to transform our data and provide us with anchors to interpret our results.

Remember that the main assumption of linear regression is the existence of a linear relationship.

Data pre-processing is nothing new in the data science workflow.

One of those techniques consists on centring the predictors such that their mean is zero, and is often used in regression analysis. Among other things it leads to interpreting the intercept term as the expected value of the target variable when the predictors are set to zero. Another useful transformation is the scaling of our variables. This is convenient in cases where we have features that have very different scales, where some variables have large values and others have very small ones.

Centring the data around their mean leads us to interpret the intercept as the expected value of the target variable when the predictors are set to zero.

As mentioned above, standardisation and scaling may help us interpret our results: They allow us to transform the features into a comparable metric with a known range, mean, units and/or standard deviation. It is important to note that the transformations to be used depend on the dataset and the domain where the data is sourced from and applied to. It also depends on the type of algorithm and answer sought. For example, in a comprehensive study of standardisation for cluster analysis, Milligan and Copper[5] report that standardisation approaches that use division by the range of the feature provide a consistent recovery of clusters. We shall talk about clustering in the next chapter.

The transformations to be applied depend on the dataset and the domain of application.

[5] Milligan, Glenn W. and Cooper, Martha C. (1988). A study of standardization of variables in cluster analysis. *Journal of Classification* 5(2), 181–204

Let us go through the two techniques mentioned above in a bit more detail.

4.6.1   *Normalisation or Unit Scaling*

THE AIM OF THIS TRANSFORMATION is to convert the range of a given variable into a scale that goes from 0 to 1. Given a feature f with a range between f_{min} and f_{max} the transformation is given by:

$$f_{scaled} = \frac{f - f_{min}}{f_{max} - f_{min}}. \qquad (4.24)$$

Unit scaling transforms our data into a scale between 0 and 1.

Notice that this method of scaling will cast our features into equal ranges, but their means and standard deviations will be different. An alternative formulation divides each feature by its range without subtracting the minimum

Unit scaling leaves the means and standard deviations unchanged.

value. We can apply this unit scaling to our data with the preprocessing method in Scikit-learn that includes the MinMaxScaler function to implement unit scaling.

```
from sklearn import preprocessing
```

Scikit-learn comes with a preprocessing module.

Once we have loaded the appropriate function, we can apply the scaling as follows:

```
scaler = preprocessing.MinMaxScaler()

mammals_minmax = pd.DataFrame(\
scaler.fit_transform(mammals[['body', 'brain']]),\
columns = ['body','brain'])
```

Scikit-learn includes MinMaxScaler to implement unit scaling.

Let us see the minimum and maximum values of the transformed data:

```
> mammals_minmax.groupby(lambda idx: 0).\
agg(['min','max'])


   body        brain
   min   max   min   max
0  0.0   1.0   0.0   1.0
```

Here we are using Pandas to manipulate our dataset.

4.6.2 z-Score Scaling

AN ALTERNATIVE METHOD FOR SCALING our features consists of taking into account how far away data points are from the mean. In order to provide a comparable measure, the distance from the mean is calculated in units of the standard deviation of the feature data.

The standard deviation is a measure of dispersion.

In this case a positive score tells us that a given data point is above the mean whereas a negative one is below the mean. The standard score explained above is called the z-score as it is related to the normal distribution. The transformation that we need to carry out for a feature f with mean μ_f and standard deviation σ_f is given by:

z-score scaling is related to the normal distribution.

$$f_{z-score} = \frac{f - \mu_f}{\sigma_f}. \qquad (4.25)$$

Strictly speaking, the z-score must be calculated with the mean and standard deviation of the population, otherwise we are making use of the Student's t statistic[6].

[6] Freedman, D., R. Pisani, and R. Purves (2007). *Statistics*. International student edition. W.W. Norton & Company

Scikit-learn's preprocessing method allows us to standardise our features in a very straightforward manner:

```
scaler2 = preprocessing.StandardScaler()

mammals_std = pd.DataFrame(\
scaler2.fit_transform(mammals[['body','brain']]),\
columns = ['body','brain'])
```

Scikit-learn includes StandardScaler to implement z-score scaling.

After the transformation we should have features with zero mean and standard deviation one; let us check that this is the case:

```
> mammals_std.groupby(lambda idx: 0).\
 agg(['mean','std'])
```

Once again, the aggregation shown here uses Pandas.

	body		brain	
	mean	std	mean	std
0	1.790682e-18	1.008163	-3.223228e-17	1.008163

4.7   Polynomial Regression

IN THE PREVIOUS SECTION WE have seen how a simple transformation in the input and output variables make a complex model into a simpler one. In fact, we can try fitting different models using more and more complex functions. One important point to note is that a model is said to be linear when it is linear *in the parameters*. With that in mind, the model

$$y = \beta_0 + \beta_1 x + \beta_2 x^2 + \varepsilon, \qquad (4.26)$$

A model is said to be linear when it is linear in the parameters.

and the model

These two models are linear.

$$
y = \beta_0 + \beta_1 x + \beta_2 x_2 + \beta_{11} x_1^2 + \\
\beta_{22} x_2^2 + \beta_{12} x_1 x_2 + \varepsilon \qquad (4.27)
$$

are both linear as the parameters β_i are linear. In the case of the examples above, the models are given by second order polynomials in one and two variables, respectively. When using such models to fit our data we talk therefore about *polynomial regression* and in general the k-th order

The models above are given by second order polynomials in one and two variables respectively. We can then talk about polynomial regression.

polynomial model in one variable is given by:

$$y = \beta_0 + \beta_1 x + \beta_2 x^2 + \cdots + \beta_k x^k + \varepsilon. \qquad (4.28)$$

This is general polynomial model.

The techniques for fitting a linear regression model can be used for the models above too.

Polynomial models can be very useful in cases where we know that nonlinear effects are present in the target variable. The polynomial model is effectively the Taylor series expansion of an unknown function and thus can be used to approximate it. Furthermore, it is possible to use different orthogonal functions to define the model. For instance, if we decide to use trigonometric functions we end up effectively in the realm of harmonic analysis and the regression would give us the coefficients we would obtain via a Fourier transform.

A polynomial model is effectively the Taylor series expansion of an unknown function.

Let us fit a quadratic model to the brain and body dataset we have been using in the previous sections. We can start by adding a feature that corresponds to the square of the body mass:

```
mammals['body_squared']=mammals['body']**2
```

In this case we are trying a quadratic model with our test dataset.

We can now fit the quadratic model given by Equation (4.26), and using StatsModels is a straightforward task:

```
poly_reg=smf.ols(formula=\
    'brain~body+body_squared',\
    data=mammals).fit()
```

The application of OLS remains the same.

We can take a look at the parameters obtained:

```
> print(poly_reg.params)


Intercept      19.115299
body            2.123929
body_squared   -0.000189
```

The parameter for the quadratic term seems to be small.

In other words, we have a model given by

$$Brain = 19.115 + 2.124(Body) - 1.89 \times 10^{-4}(Body)^2. \quad (4.29)$$

It may seem that the coefficient of the quadratic term is rather small, but it does make a substantial difference to the predictions. Let us take a look by calculating the predicted values and plot them against the other two models:

```
poly_brain_pred=poly_reg.predict(exog=\
    dict(body=new_body,\
    body_squared=new_body**2))
```

However, it makes a substantial difference to the predictions made.

As we can see in Figure 4.8, the polynomial regression captures the data points much closer than the other two models. However, by increasing the complexity of our model we are running the risk of overfitting the data. We know that cross-validation is a way to avoid this problem, and other techniques are at our disposal, such as performing some feature selection by adding features one at a time (forward selection) or discarding non-significant ones (backward elimination). In Section 4.9 we shall see how feature selection can be included in the modelling stage by applying regularisation techniques.

We have to bear in mind that increasing the complexity of our model, increases the chances of overfitting.

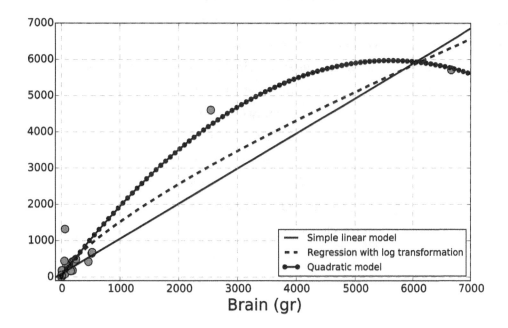

Figure 4.8: A comparison of a quadratic model, a simple linear regression model and a model with logarithmic transformation fitted to the brain (gr) versus body mass (kg) for various mammals.

When using polynomial regressions there are a number of things that should be taken into account. For instance, the order of the polynomial model should be kept as low as possible; remember that we are trying to generalise and not run an interpolation.

Once a model is obtained, take particular care not to overuse the model; extrapolating with the aid of a polynomial model is a perilous task. There are other technical issues to be aware of. For instance, as the order of the polynomial increases, matrix inversion calculations become inaccurate; this is a form of ill conditioning and it introduces errors in the estimation of the parameters.

Extrapolating with a polynomial model is a perilous task.

Another aspect to take into account is that if the values of the independent variables are limited to a narrow range, there can also be significant ill conditioning of the problem, or multicollinearity in the features used to train the model. Multicollinearity is the name we give to the situation where two or more features in our model are highly (or even moderately) correlated with each other. This becomes aggravated in a polynomial regression as higher powers of a feature are highly correlated with each other. Let us look at this empirically by taking the 9th and 10th powers of an array and calculating their correlation:

> Multicollinearity arises when two or more features are highly correlated with each other.

```
x = np.random.random_sample(500)


x1, x2 = x**9, x**10


cor_mat = np.corrcoef(x1,x2)
```

Let us take a look at the correlation matrix:

```
> print(cor_mat)


[[ 1.          0.99877083]
 [ 0.99877083  1.         ]]
```

> As we can see, the correlation between x^9 and x^{10} is high.

As we can see for the random numbers generated by my computer, the correlation coefficient between x1 and x2 is quite close to 1. Multicollinearity results in having wide swings in the values of the parameter estimations when small changes in the data are included. Also, the coefficients obtained may be such that their standard errors are quite

> The numbers may be different in your computer.

high with low significance levels, although they are actually jointly significant and the R^2 is high.

It is important to note that multicollinearity is not exclusive of the use of polynomial models. In fact, it is quite possible that two seemingly independent features included in our data are highly correlated among themselves, having a confounding effect in our model and thus we should avoid using these features together in our model.

Remember that multicollinearity is not exclusive of polynomial models.

4.7.1   Multivariate Regression

IN THE EXAMPLES SO FAR, we have concentrated mainly on regression models that have one single independent variable to explain the target we are interested in. As mentioned earlier in this chapter, in the case where we have more than one input variable we are entering the realm of multivariate regression.

Multivariate regression refers to having more than one input variable in our model.

In a sense we have already - indirectly - seen an example of a multivariate regression in Section 4.7 when we addressed the ideas behind polynomial regression. In that case, the added features were powers of a single input variable. For the more general case of a multivariate regression the features are given by different independent variables.

Polynomial regression is effectively a multivariate regression problem.

If we consider a set of M predictors $x_1, x_2, x_3, \ldots, x_M$ that are hypothesised to be related to a response variable y, the multivariate regression model can be expressed as:

$$y = \beta_0 + \beta_1 x_1 + \beta_2 x_2 + \cdots + \beta_r x_M + \varepsilon, \qquad (4.30)$$

A general multivariate linear model.

and the best part is that the parameter estimation for this
model can be achieved with the same techniques discussed
in Section 4.3. This means that we can continue using the
same StatsModels libraries described in the previous
sections. Also, as before, multicollinearity should be
avoided when considering the various independent features
to be included in our model.

We can continue using
StatsModels for polynomial
regression.

4.8   Variance-Bias Trade-Off

Now that we have explored the ideas behind describing
the relationship among variables with a model of the type
given the expression:

$$y = f(x) + \varepsilon, \tag{4.31}$$

we can take a look at the expected prediction error obtained
when estimating a model $\hat{f}(x)$. This is given by the
expectation of the squared error

$$E\left[\left(y - \hat{f}(x)\right)^2\right]. \tag{4.32}$$

The expectation of the squared
error can be decomposed into bias,
variance and noise.

This expectation value can be can be decomposed into
portions that correspond to bias, variance and noise
respectively.

In order to facilitate the decomposition let us first consider a
random variable Z with a probability distribution given by
$P(Z)$. We denote the expectation value of Z as $E[Z]$. Let us
calculate the expectation value of $(Z - E[Z])^2$:

$$E\left[(Z - E[Z])^2\right] = E\left[Z^2\right] - 2E[Z]E[Z] + E^2[Z],$$

$$= E\left[Z^2\right] - 2E^2[Z] + E^2[Z],$$

We will use the result of this calculation later on.

$$= E\left[Z^2\right] - E^2[Z]$$

and thus

$$E[Z^2] = E[(Z - E[Z])^2] + E^2[Z]. \qquad (4.33)$$

Using the expression above, we can now take a look at the decomposition of the expectation of the squared error:

$$E\left[\left(y - \hat{f}(x)\right)^2\right] = E\left[y^2 - 2\hat{f}(x) + \hat{f}^2(x)\right],$$

$$= E\left[y^2\right] - 2E[y]E\left[\hat{f}(x)\right] + E\left[\hat{f}^2(x)\right],$$

$$= E\left[(y - E[y])^2\right] + E^2[y]$$
$$-2E[y]E\left[\hat{f}(x)\right]$$
$$+E\left[\left(\hat{f}(x) - E\left[\hat{f}(x)\right]\right)^2\right]$$
$$+E^2\left[\hat{f}(x)\right]$$

$$= E\left[\left(\hat{f}(x) - E\left[\hat{f}(x)\right]\right)^2\right] + \qquad (4.34) \quad \text{Variance}$$
$$\left(E[y] - E\left[\hat{f}(x)\right]\right)^2 + \qquad (4.35) \quad \text{Bias}$$
$$E\left[(y - E[y])^2\right] \qquad (4.36) \quad \text{Noise}$$

where the first term (4.34) corresponds to **variance**, the second one (4.35) to the square of the **bias** and finally the third (4.36) is the **noise**.

This decomposition shows that apart from the noise, there are two sources of error in our model. Our task is the minimisation of these two error sources that preclude our algorithm from generalising.

We need to find a balance between the variance and bias.

On the one hand variance tells us how sensitive the model is to small fluctuations in the training set, on the other hand bias is related to the difference between the expected value of our estimator and its true value. High variance results in overfitting whereas high bias results in under-fitting. Finding a good model is therefore a matter of balancing the bias and the variance. This tradeoff applies to algorithms used in supervised learning.

High variance gives us more complex models, whereas high bias yields simpler ones.

4.9   Shrinkage: LASSO and Ridge

THE DECOMPOSITION OF OUR PREDICTION error into its variance and bias components makes it clear that a balance between the two is required for any regression problem we may encounter. In general, linear regression exhibits high variance and low bias and it should therefore stand to reason that lowering the variance at the expense of the bias is the way to go.

Linear regression exhibits high variance and low bias.

Furthermore, we have also seen that our ability to interpret the outcome when adding more and more features is diminished. It would be therefore preferable to identify

Feature selection is not possible within the straightforward linear model.

those features that are deemed to be the most important ones. Unfortunately, our linear regression model, as it stands at the moment, does not allow us to do this automatically.

Let us recall that given our model

$$y_i = \beta_0 + \sum_{j=1}^{M} \beta_j \mathbf{x}_j + \varepsilon_i, \qquad (4.37)$$

This is the linear model which we have been using all along.

we are interested in choosing the coefficients β_0 and β_j in order to minimise the Ordinary Least Squares (OLS) criterion given by Equation (4.8) which is just the sum of squared errors. The coefficients are in effect a way of determining whether a particular feature is important or not. In particular, the closer the coefficient is to zero the less significant the feature is.

The model coefficients are a way to determine if a feature is important or not.

Let us then consider replacing the estimates with a smaller value such that

$$\tilde{\beta}_k = \frac{1}{1+\lambda} \beta_k. \qquad (4.38)$$

The parameter λ lets us tune the value of the coefficients.

When $\lambda = 0$ our coefficients are unchanged, and as λ gets larger and larger, the coefficients start shrinking down to zero. In this manner, with the right choice of λ we can get an estimator with an improved error. The estimate is biased, but remember that we were happy to sacrifice some of that to make up for the variance.

The new estimates are biased though.

Shrinkage of the coefficients is therefore a form of regularisation as we penalise the model for increased complexity as given by the size of the coefficients. In Section 3.8 we introduced the $L2$- and $L1$-norms and it is thus natural to consider these measures for the size of the coefficients.

The use of the $L2$-norm results in the so-called **ridge regression**[7]:

$$\hat{\beta}^{ridge} = \min\left\{|\mathbf{Y} - \mathbf{X}\beta|_2^2 + \lambda\,|\beta|_2^2\right\}, \qquad (4.39)$$

whereas the application of the $L1$-norm leads to the Least Absolute Shrinkage and Selection Operator or **LASSO**[8] for short:

$$\hat{\beta}^{lasso} = \min\left\{|\mathbf{Y} - \mathbf{X}\beta|_2^2 + \lambda\,|\beta|_1\right\}. \qquad (4.40)$$

In both cases, as the value of λ is increased, the bias increases, whereas the variance decreases. It controls the amount of penalty imposed on the model and therefore it is important to find a good value for this parameter. **Model selection** is the process of finding the appropriate value for the hyperparameter, and cross-validation is a good way to tackle this problem.

For instance, using k-fold cross-validation and a set of possible hyperparameter values $\lambda \in \{\lambda_1, \ldots, \lambda_m\}$, we partition our data in K folds: $F_1, F_2, \ldots, F_k$. For each value of $k = 1, \ldots, K$, we train on the feature values in the training set F_i with $i \neq k$ and validate on the feature values in F_k.

For each value in the set λ we compute our estimate on the training set as well as the error on the validation set and

[7] Hoerl, A. E. and R. W. Kennard (1970). Ridge regression: Biased estimation for nonorthogonal problems. *Technometrics* 12(3), 55–67

[8] Tibshirani, R. (1996). Regression Shrinkage and Selection via the Lasso. *J. R. Statist. Soc. B* 58(1), 267–288

The tuning parameter λ is sometimes called the hyperparameter.

We discussed cross-validation in Section 3.12.

A quick recipe to find a suitable value for the hyperparameter λ, using cross-validation.

compute the average error over all folds. The latter provides us with a curve that corresponds to the cross-validation error. The value of the hyperparameter to choose is such that it minimises the cross-validation error itself which in turns corresponds to the best score for the model.

Do not be deceived by the similar look between the ridge and lasso regressions; the solutions do have significant differences. Although the ridge regression works well in cases where there are coefficients whose values are actually close to zero, the algorithm never explicitly sets them to this value.

Ridge and LASSO differ on the type of penalty imposed.

Unless $\lambda = \infty$.

This means that in some cases, with ridge regression the feature selection wished for is not possible, particularly when there are a large number of features involved. In the case of LASSO, the usage of the $L1$-norm as the penalty means that it is possible for some of the coefficients to be shrunk down to zero, making feature selection possible.

Feature selection is possible with LASSO.

It is important to remember that if the features included in our model are not on the same scale, the estimates obtained with both ridge and LASSO are not fair. In those cases it is recommended to use some scaling as described in Section 4.6. Let us know see how we can run ridge and LASSO regressions in Python. In this case we are going to use Scikit-learn to carry out the modelling.

We discussed scaling in Section 4.6.

We will continue working with the body and brain dataset we have been using all along, and in order to make things more interesting we will use a feature that corresponds to the cube of the body size. We can do this as follows:

Assuming that we have already added the square of the bodysize.

```
mammals['body_cubed']=mammals['body']**3
```

A contrived feature, but it will
serve our purposes for a demo.

Let us first start by scaling our data using z-score scaling:

```
from sklearn import preprocessing
X = mammals[['body','body_squared','body_cubed']]
Y = mammals[['brain']]


Xscaled = preprocessing.\
StandardScaler().fit_transform(X)
Yscaled = preprocessing.\
StandardScaler().fit_transform(Y)
```

As we know, this scaling can be
done with StandardScaler.

Not only are we interested in finding the coefficients that
describe each of the ridge and LASSO models, but we also
want to find a good value for λ in each case. In order to do
that we will need to carry out the cross-validation procedure
as described earlier in this section.

Fortunately, Scikit-learn provides us with GridSearchCV
which is a helpful function that lets us perform an
exhaustive search over specified parameter values by
implementing a fit and a score methods. The latter will let
us choose the value for our hyperparameter λ. First, we
need to create our test and training sets:

Scikit-learn lets us carry
out an exhaustive search of
parameter combinations with the
GridSearchCV method.

```
import sklearn.model_selection as ms


XTrain, XTest, yTrain, yTest =\
ms.train_test_split(Xscaled, Yscaled,\
test_size= 0.2, random_state=42)
```

We split our data into training and
testing sets.

We can now define a set of parameters to be used in our search:

```
from sklearn.model_selection import GridSearchCV
from sklearn.linear_model import Ridge, Lasso


lambda_range = linspace(0.001,0.2,25)
lambda_grid = [{'alpha': lambda_range}]
```

We need to define a dictionary that holds the set of values to be searched. Note that Scikit-learn refers to the hyperparameter as α.

Our search will use each of the values in the `lambda_grid` dictionary and carry our cross-validation with the number of folds desired:

```
model1 = Ridge(max_iter=10000)
cv_ridge = GridSearchCV(estimator=model1,\
param_grid=lambda_grid,\
cv=ms.KFold(n_splits=20))
cv_ridge.fit(XTrain, yTrain)


model2 = Lasso(max_iter=10000)
cv_lasso = GridSearchCV(estimator=model2,\
param_grid=lambda_grid,\
cv=ms.KFold(n_splits=20))
cv_lasso.fit(XTrain, yTrain)
```

GridSearchCV takes care of the cross-validation step using the set of parameters to be searched. In this case we are using k-fold cross-validation.

In Figure 4.9 we show a heatmap with the values for the hyperparameters used in our search and their method, as well as their corresponding cross-validation mean scores. We can obtain the actual values selected with the help of the `best_params_` method as follows:

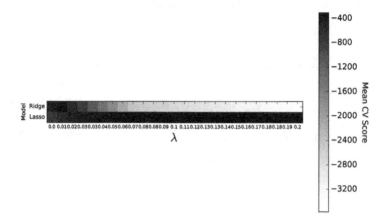

Figure 4.9: Using `GridSearchCV` we can scan a set of parameters to be used in conjunction with cross-validation. In this case we show the values of λ used to fit a ridge and LASSO models, together with the mean scores obtained during modelling.

```
> cv_ridge.best_params_['alpha'],\
  cv_lasso.best_params_['alpha']


(0.133666666667, 0.00929166666667)
```

We can now use these parameters with the corresponding models to extract the coefficients:

```
> bestLambda_lasso=cv_lasso.best_params_['alpha']
> Brain_Lasso = Lasso(alpha=bestLambda_lasso,\
  max_iter=10000)
> Brain_Lasso.fit(XTrain,yTrain)
> print(Brain_Lasso.coef_)
[ 1.65028091 -0.          -0.76712502]
```

Similar code can be written for the ridge model.

as we can see, the second coefficient has been shrunk down to zero with the application of the LASSO regression.

Finally, let us take a look at the residual sum of squares obtained with the test dataset:

```
> lasso_prediction = Brain_Lasso.predict(XTest)
> print(''Residual sum of squares:%.4f ''\
        % np.mean((lasso_prediction - yTest)**2))


Residual sum of squares: 0.0114
```

Once again, similar code can be written for the ridge model.

4.10 Summary

IN THIS CHAPTER WE HAVE addressed the topic of regression analysis. It is a natural first step in our data science and analytics journey as it is one of the most widely used techniques out there. Mastering regression is a must for a jackalope data scientist.

We have seen how regression allows us to describe relationships between input features and the target variable, bearing in mind that correlation does not imply causation.

Using the language of linear algebra we implemented the Ordinary Least Squares (OLS) model to solve the linear regression problem and extended this to a multivariate situation. Furthermore, we have also seen how the models can be used to carry out polynomial regression.

The use of appropriate transformations to our input (and output) data was shown to be beneficial to our modelling task, and the interplay between bias and variance in our models is an important concept to take into account at the modelling stage.

The continuous tension between bias and variance can be used to our advantage in the form of regularisation techniques such as ridge and LASSO, allowing us to fine tune our models in a very flexible manner. Regression is indeed a useful tool that every jackalope data scientist should have in the toolbox.

5

Jackalopes and Hares: Clustering

HAVE YOU EVER CONSIDERED HOW we can distinguish
a rabbit from a stag, or a jackalope from a hare? Since we
are able to do this instantaneously, this seems to be a very
silly question. A jackalope is effectively the same as a hare
except for the antlers. In the case of the rabbit and the stag,
if the animal is small and has long ears then it is a rabbit,
whereas if it has prominent antlers then it is a stag. We use
distinguishing features that allow us to create groups based
on similarities and differences.

Of course, if we come across a
jackalope we may have to think
twice.

We talk about **clustering** when the groups made out of
similar data points do not have a predefined name or label.
When the label does exist we talk about **classification** and
will cover it in Chapter 6. Clustering analysis is an
unsupervised machine learning task, whereas classification
is a supervised one. In this chapter we will present some
important algorithms that enable us to cluster hares and
jackalopes, as well as rabbits and stags.

Clustering is an unsupervised
learning task, whereas
classification is a supervised
one.

5.1   *Clustering*

A CLUSTER CAN BE THOUGHT of as a group of similar data points and therefore the concept of similarity is at the heart of the definition of a cluster. The greater the similarity among points leads to better clustering and thus to better results. We have mentioned that clustering analysis is an unsupervised learning task. This means that its goal is to provide us with a better understanding of our dataset by dividing the data points into relevant groups. Once the clusters are defined, we can assign them a label and use them as the starting point for classification of unseen data.

A cluster is a group of similar data points, and the concept of similarity is therefore important.

Let us consider a couple of examples: Imagine an alien life-form arriving to Earth and being presented with the animals listed in the dataset used in Chapter 4. It does not know what a cat, a rabbit, a horse and a roe deer are and it does not even have a name for these strange-looking animals.

Imagine an alien life-form that does not know the animals on Earth.

The alien is shown examples of various specimens, and starts to make notes: Cats have pointy triangular ears, whereas rabbits have long oval ones; horses have manes and deer have antlers. In other words, our alien friend is looking at the similarities (and differences) among the specimens presented for examination. This enables it to group the animals even if it does not know what humans call them. With the help of its clusters, our friendly alien can create its own labels for the groupings. The next time the alien sees a small mammal with a round head, whiskers and triangular ears, it will be able to classify this animal with other cats.

Our alien friend when presented with a variety of animals can cluster them by their similarities, even if it does not know the names we give to these animals.

The application of clustering in the simple example above shows how it helps us enhance our knowledge of the dataset we are working with. Clustering provides us with a layer of abstraction from individual data points to collections of them that share similar characteristics. It is important to clarify that the enhancement is made by extracting information from the inherent structure of the data itself, rather than imposing an arbitrary external one.

In that sense, a cluster can be conceived to be a *potential* class, and the solution to a clustering problem is determining these classes, as per the data at hand. Clustering is therefore a data exploration technique that contributes to our familiarity with the dataset. Once we have created suitable clusters and provided them with a label, these clusters can be the starting point to use supervised machine learning techniques. This is true not only about clustering, but also about other unsupervised learning algorithms we will encounter in the rest of the book.

Clustering enables us to enhance our knowledge of our datasets.

A cluster is a potential class.

Clustering is a data exploration technique.

5.2   *Clustering with k-means*

ONE OF THE SIMPLEST ALGORITHMS that one can implement to solve a clustering problem is the so-called k-means algorithm[1]. Its goal is to partition an N-dimensional dataset into k different sets, whose number is fixed at the start of the process. The algorithm performs a complete clustering of the dataset, in other words, each data point considered will belong to exactly one of the k clusters.

[1] MacQueen, J. (1967). Some Methods for classification and Analysis of Multivariate Observations. In *Proceedings of 5-th Berkeley Symposium on Mathematical Statistics and Probability*. University of California Press

The k-means procedure is said to be a *greedy* algorithm as it employs a heuristic based on local optimal choices and therefore the solution found depends on the initial conditions given. The most important part of the process is determining the partitions that form the k sets. This is done by defining k centroids and assigning each data point to the cluster with the nearest centroid. The centroid is then updated by taking the mean of the data points in the cluster.

A heuristic helps us solve a problem in a quick fashion by finding an approximate solution.

k-means requires the definition of k partitions to which data instances are assigned.

From the brief description above it is clear that the data required needs to have features in a vector-like form. We also have to bear in mind that the process is iterative in nature. Two other important things to consider are the following:

- The partitions are not scale-invariant and therefore the same dataset may lead to very different results depending on the scale and units used. In Figure 5.1 we show two representations of a same dataset using two scales

The partitions are not scale-invariant.

- The initial k centroids are set at the beginning of the process and different locations may lead to different results

Different k centroids lead to different results.

Remember that the number of clusters is set at the beginning of the modelling process. The general idea behind k-means can be summarised in the following four steps:

The value k is an input to the algorithm.

1. Choose the location of the initial k centroids

2. For each data point, find the distance to each of the k centroids and assign the point to the nearest one

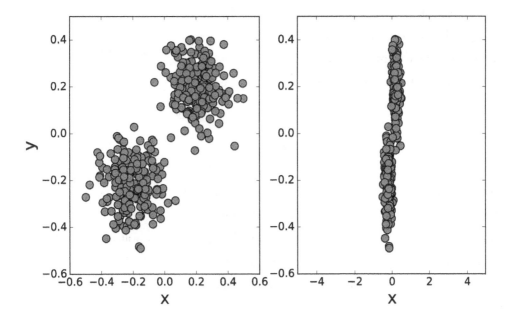

3. Once all data points have been assigned to a cluster, recalculate the centroid positions

4. Repeat steps 2 and 3 until convergence is met

Step 1 above requires us to choose the initial centroids and there are several possibilities for this. One can choose the positions at random. We can also start with a global centroid and choose points at a maximum distance from it. A good alternative is using multiple random initial conditions on various clustering trials.

We know that a cluster is defined by similarity among data points. In the case of k-means that similarity is shown by the closeness of a data point to a given centroid. That closeness

Figure 5.1: The plots show the exact same dataset but in different scales. The panel on the left shows two potential clusters, whereas in the panel on the right the data may be grouped into one. The centroids will no longer change position.

Note that a random choice may lead to a divergent behaviour.

is determined by the distance measure or similarity measure chosen for the task. We know the conditions that should be met by a good similarity measure as discussed in Section 3.8. An intuitive way to measure similarity is the Euclidean distance calculated from the N features that describe the data points.

Similarity among points is measured by their closeness to the centroids.

As is the case with many algorithms, k-means aims at minimising an objective function. The optimisation of this objective function lets us know how well our clustering task is performing. The recalculation of the centroids at each iteration has to be carried out with this in mind. Using the Euclidean distance $d(x, c_i)$ from point x to the centroid c_i, a typical objective function is a squared error function that is given by:

As with other algorithms, k-means aims at minimising an objective function too.

$$SSE_{k-means} = \sum_{i=1}^{k} \sum_{i}^{N} d\left(x_j^{(i)}, c_i\right)^2. \tag{5.1}$$

As such, given two clustering results, we will prefer the one with the lower sum of squared errors. This is an indication that the centroids have converged to better locations and therefore to a better local optimum for the objective function above.

5.2.1   Cluster Validation

IT IS IMPORTANT TO NOTE that even in cases where no partion exists, k-means will return a partition of the dataset in to k subsets. It is therefore useful to validate the clusters obtained. Cluster validation can be further used to identify

Cluster validation is an important part of the process to determine the effectiveness of the algorithm.

clusters that should be split or merged, or to identify
individual points with disproportionate effect on the overall
clustering.

This can be done with the help of two measures: Cohesion
and separation. **Cohesion** is a measure of how closely
related data points within a cluster are, and is given by the
within-cluster SSE:

$$C(c_i) = \sum_{x \in c_i} d(x, c_i)^2. \tag{5.2}$$

Cohesion tells us how closely
related the data points are in a
given cluster.

Separation is a measure of how well clusters are segregated
from each other:

$$S(c_i, c_j) = d(c_i, c_j)^2. \tag{5.3}$$

Separation indicates if clusters are
self-contained.

Figure 5.2 shows a diagrammatic representation of cluster
cohesion and separation. The overal cohesion and
separation measures are given by the sum over clusters; in
the case of separation it is not unusual to weight each of the
terms in the sum.

We can use these two definitions to provide us with an
overall measure of clustering validity $V_{overall}$ by taking a
weighted sum over clusters such that

$$V_{overall} = \sum_{i=1}^{k} w_i V(C_i), \tag{5.4}$$

The overall clustering validity
measure combines both cohesion
and separation.

where V can be cohesion, separation or a combination of
them.

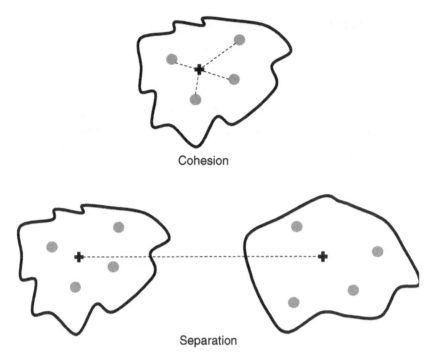

Cohesion

Separation

Figure 5.2: A diagrammatic representation of cluster cohesion and separation.

An alternative measure of validity that provides us with a combination of the ideas behind cohesion and separation in a single coefficient is given by the **silhouette**[2] score.

[2] Rousseeuw, P. J. (1987). Silhouettes: a Graphical Aid to the Interpretation and Validation of Cluster Analysis. *Comp. and App. Mathematics 20*, 53–65

For a point x_i the silhouette is defined by the average in-cluster distance to x_i, denoted by a_i, and the average between-cluster distance to x_i, denoted by b_{ij}. Out of these measures it is possible to obtain $b_i = \min_j(b_{ij})$:

$$s(x_i) = \frac{b_i - a_i}{\max(a_i, b_i)}, \tag{5.5}$$

The silhouette coefficient combines the ideas behind cohesion and separation.

and it has values between -1 and 1. The value of a_i tells us how dissimilar x_i is in the cluster and thus we would prefer small values for this quantity. Having a large b_i implies that x_i has been badly matched to a cluster nearby. We are interested in having high separation and low cohesion and this situation corresponds to values close to 1 for the silhouette coefficient. The average silhouette over the entire dataset tells us how well the clustering algorithm has performed and can be used to determine the best number of clusters for the dataset at hand.

All in all, k-means is pretty efficient both in time and complexity, however it does not perform very well with non-convex clusters, or with data having varying shapes and densities. One possible way to deal with some of these issues is by increasing the value of k, and later recombining the sub-clusters obtained. Also, remember that k-means requires a carefully chosen distance measure that captures the properties of the dataset.

5.2.2   k-means in Action

Let us take a look at an example of clustering with k-means using Scikit-learn. We will use data that record the results of chemical analysis of Italian wines grown in the same region and from three different cultivars. The dataset can be found in the UCI Machine Learning Repository under "Wine Dataset"[3] and is available at http://archive.ics. uci.edu/ml/datasets/Wine.

The 13 attributes determined by the chemical analysis are: Alcohol, Malic acid, Ash, Alcalinity of ash, Magnesium, Total phenols, Flavonoids, Nonflavonoid phenols, Proanthocyanins, Colour intensity, Hue, OD280/OD315 of diluted wines, and Proline. The data also has information

We are interested in having high separation and low cohesion.

Complexity is linear in the number of records.

[3] Lichman, M. (2013a). UCI Machine Learning Repository, Wine Data. https://archive. ics.uci.edu/ml/datasets/ Wine. University of California, Irvine, School of Information and Computer Sciences

about the cultivar where the wine is from. In this case we will not use this information as *k*-means is an unsupervised machine learning algorithm.

We have pre-processed the data into a csv file with column names and we can now used pandas to read the file:

We have preprocessed the data to add appropriate column names.

```
wine = pd.read_csv(u'./Data/wine.csv')
```

We can see the names of the columns in our dataset:

```
> wine.columns

Index([u'Cultivar', u'Alcohol', u'Malic_Acid',
       u'Ash', u'Ash_Alcalinity', u'Magnesium',
       u'Total_Phenols', u'Flavonoids',
       u'NonFlavonoid_Phenols',
       u'Proanthocyanins', u'Colour_Intensity',
       u'Hue', u'0D280_0D315_DilutedWines',
       u'Proline'],
      dtype='object')
```

These are the features included in the wine dataset.

In order to simplify the example, we are going to concentrate on a couple of features out of the 13 mentioned above. In this situation, we will use the Alcohol and Colour Intensity features as they are some of the more obvious descriptors that a wine tester would note. We will therefore create an array that contains the values of these columns and while we are at it we will also extract the cultivar for each of the wines:

Perhaps a wine chemist or a proper sommelier would start with other features.

```
X1=wine[['Alcohol','Colour_Intensity']].values
Y=wine['Cultivar'].values
```

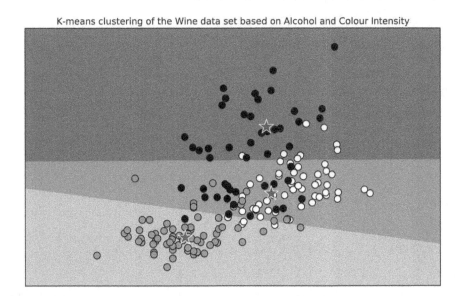

K-means clustering of the Wine data set based on Alcohol and Colour Intensity

Figure 5.3: *k*-means clustering of the wine dataset based on Alcohol and Colour Intensity. The shading areas correspond to the clusters obtained. The stars indicate the position of the final centroids.

We will use $k = 3$ as we know that there are three different cultivars in the dataset. In general, we would not necessarily have this information to start with. In that case a search of the optimal k can be carried out with the help of the silhouette score. Here we will use the KMeans method in Scikit-learn's cluster:

```
from sklearn import cluster
cls_wine = cluster.KMeans(n_clusters = 3)
cls_wine.fit(X1)
```

The method KMeans takes the parameter n_clusters to indicate the number of clusters k.

The shading areas shown in Figure 5.3 illustrate the clusters obtained for the wines in the dataset using the alcohol and colour intensity features. The data points are shown as filled circles and their colours correspond to their actual cultivar as per the dataset itself. Finally, we mark the position of the final centroids with the help of filled stars.

We have used $k = 3$ in the example above and only used a couple of the features available.

We can take a look at the clusters to which each of the data points have been assigned to with the aid of labels_:

```
> print(cls_wine.labels_)

[2 2 2 0 2 2 2 2 0 2 2 2 0 0 2 2 0 2 2 2 2
 1 1 1 2 2 2 2 2 2 2 2 2 2 2 1 2 2 2 2 2 2
 2 2 2 0 0 2 0 2 2 2 2 2 1 1 2 2 2 1 2 2 2 1
 1 1 2 1 1 1 2 1 1 1 1 1 2 1 1 1 1 1 1 1
 1 1 1 1 1 2 1 1 1 1 1 1 1 1 1 1 1 1 1 1
 1 1 1 1 1 1 2 1 1 1 1 1 1 1 1 2 2 2 2 0 1 2
 2 2 2 2 2 0 2 2 0 0 0 0 0 0 0 0 0 0 0 0 0
 2 2 2 0 2 0 0 0 0 2 0 0 0 0 0 0 0]
```

The predicted labels are obtained with the .labels_ method.

The values correspond to the cluster (0, 1, or 2) in which the records have been placed. The algorithm does not know anything about the cultivars, so these numbers make no reference to the values that appear in the dataset.

The output of fitting KMeans also provides information about the location of the final centroids and it is saved in cluster_centers_:

```
> print(cls_wine.cluster_centers_)


[[ 13.38472222    8.74611108]
 [ 12.25353846    2.854      ]
 [ 13.45168831    5.19441558]]
```

The coordinates of the final centroids are obtained with the method .cluster_centers_.

Finally, we can see the silhouette score obtained with the use of the features and value of k chosen:

```
> from sklearn.metrics import silhouette_score
> print(silhouette_score(X1, cls_wine.labels_))


0.509726787258
```

The silhouette score is calculated with silhouette_score().

Remember that we only used two features. Perhaps the score can be improved either by using more relevant features or, in the case where we do not know k, by varying the number of clusters. We leave this as an exercise for the reader.

5.3 Summary

CLUSTERING IS A VERY IMPORTANT application in the data science workflow as it provide ways to explore our data. In this chapter we discussed the difference between clustering and classification. The former is an unsupervised learning task that allows us to establish meaningful groupings within our data. The latter is a supervised learning task that exploits information about the classes contained in our data

in order to make predictions about unseen data instances.
We will discuss classification in the following chapter.

We discussed the k-means algorithm for clustering data,
where data instances are assigned to their closest centroid
among the k predefined ones. We saw how the cohesion,
separation and silhouette measures let us validate the
clusters we obtain with the algorithm.

Remember that we require prior knowledge of k or
alternatively a search for an optimal value. It is worth
pointing our that a typical k-means algorithm is started with
a random choice for the initial centroids. This means that it
is possible to obtain different results on different runs of the
algorithm with the same data. This is an important point as
the clusters we obtain will need to be evaluated not just
using cohesion, separation and silhouette scores, but also for
fitness to our end goal within the business or research
context where we are employing the algorithm.

6

Unicorns and Horses: Classification

DISTINGUISHING UNICORNS FROM HORSES OR jackalopes from hares by using their differences and similarities can be very useful as we saw in the previous chapter. The groupings we can create based on the chosen features can tell us something about the members in each of them. Furthermore, depending on the context in which we create those groups or clusters we can exploit research or business knowledge and assign a label to each of the clusters.

Once we are in possession of labelled data, we can take a step further and use those labels in a supervised learning task, where the labels become our targets. In this chapter we will discuss how **classification** algorithms are used and scored. In particular we will cover some important algorithms such as K Nearest Neighbours, Logistic Regression and the famous Naïve Bayes classifier, letting us classify unicorns and horses, jackalopes and hares and why not, rabbits and stags too.

6.1   Classification

CLASSIFICATION IS A TASK THAT involves arranging
objects systematically into appropriate groups or categories
depending on the characteristics that define such groupings.
It is important to emphasise that the groups are pre-defined
according to established criteria. In our case, the use of
classification is to determine the category to which an
unseen observation belongs, depending on the information
of a training dataset with appropriate labels. Classification
is therefore a supervised learning task.

Classification is a supervised
learning task.

In Section 5.1 we encountered our alien friend trying to
make sense out of the fluffy earthlings it encountered.
Let us see another example: Imagine that we are given a
Whizzo Quality Assortment chocolate box. We have never
tried this brand, but we know that in a box of assorted
chocolates there are some delicious ones and others that
are not so good. We try a few of the sweets and notice that
among the traditional pralines there are some extremely
nasty confections: One of them has a raw frog inside, others
have cockroaches.

You may want to avoid some
of the chocolates in the Whizzo
Quality Assortment box.

We notice the shape, aroma and general look of these two
nasty sweets to make a note about them for future reference.
We even go as far as giving them a name based on the
defining attributes we have noted. In other words, we use
the features of the confections to cluster them appropriately
and provide us with a better understanding of the contents
of the box, and even label them. In that way, the next time

In particular, the "crunchy frog"
and the "cockroach cluster" may
not be very palatable.

we get another one of these chocolate boxes we know we are better off avoiding the "crunchy frog" and the "cockroach cluster" sweeties as we are able to classify them as unpalatable confections, and devour the rest of the box.

We can see the clear relationship between classification and clustering as discussed in Section 5.1. Whereas in clustering the aim is to determine the groups from the features in the dataset, classification uses the labelled groups to predict the best category for unseen data. In the context of the examples we presented earlier on, the alien life-form is able to determine whether the next animal it sees is classified to be a cat or a rabbit depending on the pre-determined groups it established at an earlier step. Similarly, we will be able to avoid eating "crunchy frogs" from our box of chocolates without the need of tasting these exotic delicacies again.

Our friendly alien life-form will be able to classify animals after having applied clustering to the specimens it was presented with.

A typical example of a classification problem is determining if a given email that arrives to an email box is *spam* or *ham*. We all know how annoying it is to get unsolicited messages that advertise for all sort of things and have no relevance to us. It would be preferable to have an email box without spam, and as such many modern email clients have implementations of classifiers whose task is to predict if an email is spam or not, and filter those messages that are deemed to be spam.

Ham of course being *not spam.*

And without *"Uuuuuuugggh!"*

These classifiers may or may not be all that great. However, they certainly try to get better at their task if they encounter more and more examples of labelled data. In the case of spam, every time that a spam message makes it through to our inbox we may get the option of marking or labelling the offending message as spam. In that way we are adding data to the training of the classifier and helping its improvement.

We help improving a spam classifier by marking misclassified emails as spam, or a legitimate messages as ham.

With that in mind, it becomes obvious that there is a need to quantify how well, or how badly, a classifier is performing. If we carried out the categorisation task at random, sometimes we would classify some observations into the right category and sometimes not. We would prefer a classifier that performs better than this random classifier. At the other end of this spectrum we have a perfect classifier, which categorises all data points into the correct class every time. Since we have labelled data to train a classifier, we can - at the very least - use this information to tell us something about the performance of the classifier. There are a few common ways to present this information and we discuss them below.

As you can imagine, getting a perfect classifier may be a hard thing to achieve.

6.1.1   Confusion Matrices

A VERY CONVENIENT WAY TO evaluate the accuracy of a classifier is the use of a table that summarises the performance of our algorithm against the data provided. Karl Pearson used the name *contingency table*[1]. These days the machine learning community tends to call it a **confusion matrix** as it lets us determine if the classifier is confusing two classes by assigning observations of one class to the other. One advantage of a confusion matrix is that it can be extended to cases with more than two categories.

[1] Pearson, K (1904). On the theory of contingency and its relation to association and normal correlation. In *Mathematical Contributions to the Theory of Evolution*. London, UK: Dulau and Co.

In any case, the contingency table or confusion matrix is organised in such a way that its columns are related to the instances in a predicted category, whereas its rows refer to actual classes. To explain the use of the confusion matrix, let us consider a binary classification system implemented for example by a World War II air reconnaissance troop. Their task consists of distinguishing enemy aircraft from flocks of birds. Let us imagine that they have taken 100 measurements and created Table 6.1. The troop is of course interested in detecting correctly when an enemy plane is flying above them.

We can also transpose the table, by the way.

		Predicted Class	
		Enemy Aircraft	Flock of Birds
	Enemy Aircraft	20	4
Actual class			
	Flock of Birds	6	70

Table 6.1: A confusion matrix for an elementary binary classification system to distinguish enemy aircraft from flocks of birds.

From Table 6.1, we can see that the troop has correctly predicted 20 enemy planes. These are called the *True Positives*. Similarly, they have correctly predicted 70 flocks, and these are called the *True Negatives*.

Please note that the word "positive" relates to the class we are interested in classifying.

A *False Positive* is a case where we have incorrectly made a prediction for a positive detection. From the table we can see that the troop has predicted 6 cases as aircraft, but they turned out to be flocks. Finally, a *False Negative* is a case where we have incorrectly made a prediction for a negative detection. Again, the table lets us see that

In that sense a false positive is a case where we have incorrectly classified an observation as belonging to the class of interest.

they have predicted 4 cases to be birds, but unfortunately they were actual enemy planes. In total we have 90 correct classifications and 10 instances of misclassification. In Table 6.2 we show a diagram of the location where we would find True Positives, False Negatives, False Positives and True Negatives in a confusion matrix.

		Predicted Class	
		Class 1	Class 2
Actual class	Class 1	True Positives (*TP*)	False Negatives (*FN*)
	Class 2	False Positives (*FP*)	True Negatives (*TN*)

Table 6.2: A diagrammatic confusion matrix indicating the location of True Positives, False Negatives, False Positives and True Negatives.

The use of a confusion matrix enables us to see how well the troop has performed the classification of enemy aircraft or flocks of birds. There are also a few quantities that help us with determining the performance of our classifier. Let us define a few of them:

A confusion matrix lets us see how well our classifier performs in the identification of the class of interest.

Recall or **True Positive Rate**: It is also known as *sensitivity* or *hit rate*. It corresponds to the proportion of positive data points that are correctly classified as positive versus the total number of positive points:

The true positive rate is also known as recall or sensitivity.

$$TPR = \frac{TP}{TP + FN}. \tag{6.1}$$

The higher the True Positive Rate, the fewer negatives will be missed. In our example $TPR = 20/24 = 0.833$.

True Negative Rate or **Specificity**: It is the counterpart of the True Positive Rate as it measures the proportion of negatives that have been correctly identified. It is given by:

$$TNR = \frac{TN}{TN + FP}.$$ (6.2)

The true negative rate is also known as specificity

In the example, $TNR = 70/76 = 0.921$.

Fallout or **False Positive Rate**: It corresponds to the proportion of negative data points that are mistakenly considered as positive, with respect to all negative data points:

The false positive rate is also known as fallout.

$$FPR = \frac{FP}{FP + TN} = 1 - TNR.$$ (6.3)

In other words, the higher the FPR, the more negative data points we will have misclassified. In our example $FPR = 6/76 = 1 - 0.921 = 0.079$.

Precision or **Postitive Predictive Value**: It is the proportion of positive results that are true positive results. Positive predictive value

The precision is also known as positive predictive value.

$$PPV = \frac{TP}{TP + FP}.$$ (6.4)

In our example $PPV = 20/26 = 0.769$.

Finally, the **Accuracy** is given by the ratio of the points that have been correctly classified and the total number of data points:

The accuracy of a classifier is the ratio of correctly classified instances and the total number of data points.

$$ACC = \frac{TP + TN}{TP + FP + FN + TN},$$ (6.5)

and in our example the accuracy is $ACC = 90/100 = 0.9$.

6.1.2   ROC and AUC

THE RECEIVER OPERATOR CHARACTERISTIC OR ROC is a
quantitative analysis technique used in binary classification.
It has its origins in the British efforts during World Word
II to differentiate enemy aircraft from noise, using systems
such as "Chain Home"[2]. Different operators had a wide
range of skills, and changes in the radar receiver gain levels
could influence signal to noise ratios. As such, flocks of
birds could be mistaken for enemy aircraft. Each receiver
operator was said to have their own characteristic and thus
the name.

Hence the example in the previous section.

[2] Galati, G. (2015). *100 Years of Radar*. Springer International Publishing

As we saw in the Section 6.1.1, we can obtain a lot of useful
information from true positives, false negatives, etc.
Nonetheless, it is sometimes easier to compare one single
metric rather than several. That is where ROC[3] becomes
very handy: It lets us construct a curve in terms of the true
positive rate against the false positive rate. Unfortunately
ROC curves are suitable for binary classification problems
only.

[3] Fawcett, T. (2006). An introduction to ROC analysis. *Patt. Recog. Lett. 27*, 861–874

In a ROC curve the **True Positive Rate** is plotted as a
function of the **False Positive Rate** for different cut-off
points or *thresholds* for the classifier. Think of these
thresholds as settings in the receiver used by the radar
operators.

The receiver operator characteristic (ROC) curve shows the true positive rate as a function of the false positive rate.

If our classifier is able to distinguish the two classes without
overlap, then the ROC would have a point at the 100%
sensitivity and 0% fallout, i.e. the upper left corner of the

A perfect classifier would have a 100% sensitivity and 0% fallout.

curve. This means that the closer the ROC curve is to that corner, then the better the accuracy of the classifier. A different way of representing this information is in terms of the area under the ROC curve, aka AUC. This means that for a perfect classifier the area under the ROC curve will be equal to 1. In that sense the AUC is the one single metric we were looking for.

A perfect classifier would have an area under the ROC curve equal to 1.

The other side of the coin is a classifier that by chance distinguishes the two classes. It would be no better than tossing a coin to assign observations to each of the two categories. This would be represented by a diagonal line from the origin and going to the $(1,1)$ point in the plot. In that case the AUC would then be 0.5. Let us see an example by considering a receiver that can be placed in 8 different settings from off to full detection. The readings from this thought experiment are as shown in Table 6.3.

A classifier that is as good as guessing would have a ROC curve described by a diagonal line.

Setting	Enemy Aircraft Detected (%)	Flocks Detected (%)	Flocks Incorrectly Detected (%)
	Sensitivity	Specificity	Fallout
Off	0	100	0
S_1	18	96	4
S_2	34	95	5
S_3	58	84	16
S_4	70	78	22
S_5	88	62	38
S_6	97	25	75
Full blast	100	0	100

Table 6.3: Readings of the sensitivity, specificity and fallout for a thought experiment in a radar receiver to distinguish enemy aircraft from flocks of birds.

In Figure 6.1 we can see the ROC curve for the data in Table 6.3. We are also showing the diagonal line that represents the 0.5 AUC classifier, as well as the curve given by a perfect classifier whose AUC is equal to 1.

Figure 6.1 shows a set of typical ROCs.

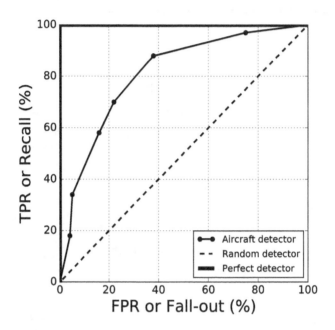

We now know what a perfect classifier looks like in terms of a ROC curve. Similarly, we also know that a diagonal line is equivalent to a random guess. With this in mind, it is clear that we would prefer classifiers that are better than guessing, in other words those whose ROC curve lies above the diagonal. Also we would prefer those classifiers whose ROC curves are closer to the curve given by the perfect classifier. If you end up with a ROC curve that lies below the diagonal, your classifier is worse than guessing, and it

Figure 6.1: ROC for our hypothetical aircraft detector. We contrast this with the result of a random detector given by the dashed line, and a perfect detector shown with the thick solid line.

We prefer classifiers that are better than guessing.

should be immediately discarded. In the following sections we will discuss some of the most popular classification algorithms and show a few examples.

6.2 Classification with KNN

NOW THAT WE HAVE CLARIFIED the goal of classification, we are in a position to discuss our first classifier: The k-Nearest-Neighbours algorithm[4] or KNN, for short. We will use the labels provided by the training data to help us decide the class to which a new unseen observation belongs.

[4] Cover, T. M. (1969). Nearest neighbor pattern classification. *IEEE Trans. Inform. Theory IT-13,* 21–27

Recall that we consider our data points to exist in an N-dimensional space given by the N features in our dataset. In the KNN classifier, similarity is given by the distance between points. We classify new observations taking into account the class of the k nearest labelled data points. This means that we need a distance measure between points, and we can start with the well-known Euclidean distance we discussed in Section 3.8.

In KNN similarity is given by the distance between data instances.

As it was the case in k-means for clustering, the value of k in KNN is a parameter that is given as an input to the algorithm. For a new unseen observation, we measure the distance to the rest of the points in the dataset and pick the k nearest points. We then simply take the most common class among these to be the class of the new observation. In terms of steps we have the following:

The number of neighbours k to take into account for classification is a parameter in this model.

1. Choose a value for k as an input

2. Select the k nearest data points to the new observation

3. Find the most common class among the k points chosen

4. Assign this class to the new observation

Please note that step 3 may pose some issues in cases where we choose an even k. For example, if $k = 4$ we may encounter a situation where two points belong to class A and the other two to class B. In this case it is not possible for us to decide whether to assign our new sample to either A or B.

An even number of neighbours may pose a decision problem, so it is better to concentrate on odd values of k.

As you can see, KNN is a very straightforward algorithm and it is very easy to explain. Furthermore, it is able to learn nonlinear boundaries among classes. However, it may be easy to overfit a dataset and the question about choosing the value of k is an important one. We also need to consider the best distance measure for our dataset. Finally, it is worth mentioning that KNN has low bias and high variance.

These issues can be dealt with with cross-validation.

6.2.1   KNN in Action

FOR THIS SECTION WE WILL make use of the Iris dataset we introduced in Section 3.10, containing 150 samples of three species of iris flowers: Setosa, Virginica and Versicolor, with four features: Sepal length, sepal width, petal length and petal width.

We introduced the Iris dataset in Section 3.10.

Let us start by creating our training and testing datasets. Remember that not only are we interested in the features, but also in the labels. We will load the features in object X and the labels in Y:

In supervised learning we are also interested in the labels.

```
X = iris.data
Y = iris.target
```

Remember to load the Iris dataset first.

With the aid of `train_test_split` we can create our testing and training datasets. In this case we are using 70% of the data for training and 30% for testing:

```
import sklearn.model_selection as ms


XTrain, XTest, YTrain, YTest =\
ms.train_test_split(X, Y,\
test_size= 0.3, random_state=7)
```

We can construct training and testing datasets with `train_test_split`. The parameter `random_state` initialises the pseudo-random number generator used for sampling.

We need to find the appropriate value of k for our problem and we can indeed fit various models for different values of k. We can facilitate this search with the help of `GridSearchCV` as we did for the hyperparameter λ in Section 4.9.

The value of k, number of neighbours, is a parameter that we need to determine.

Let us load the relevant libraries:

```
from sklearn import neighbors


from sklearn.model_selection import GridSearchCV
```

We will search odd values between 1 and 20 and find the best value of k with cross-validation:

```
k_neighbours = list(range(1,21,2))


n_grid = [{'n_neighbors': k_neighbours}]
```

We will be searching the optimal value of *k* in the odd values between 1 and 20.

We will apply this to the KNeighborsClassifier() function in the module neighbors:

```
model = neighbors.KNeighborsClassifier()


cv_knn = GridSearchCV(estimator=model,\
param_grid=n_grid,\
cv=ms.KFold(n_splits=10))


cv_knn.fit(XTrain, YTrain)
```

We are performing cross-validation with 10 folds in our training data.

The result of the search can be seen as follows:

```
> best_k = cv_knn.best_params_['n_neighbors']
> print(''The best parameter is k={0}''.\
format(best_k))


The best parameter is k=11
```

In this case we see that the best number of neighbours is 11 and in Figure 6.2 we show a heatmap of the different values of *k* considered and their corresponding score.

The score evaluated in by the model is the accuracy of the classification.

Finally, let us see how our model performs on the testing dataset using $k = 11$ for the number of neighbours. In this case we will only use two of the features of the dataset for

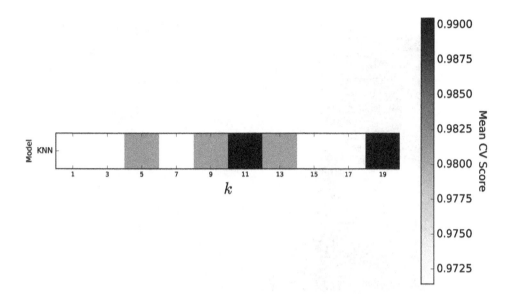

Figure 6.2: Accuracy scores for the KNN classification of the Iris dataset with different values of k. We can see that 11 neighbours is the best parameter found.

visualisation purposes. We will concentrate on the sepal width and the petal length:

```
knnclf = neighbors.KNeighborsClassifier\
(n_neighbors=best_k)

knnclf.fit(XTrain[:, 2:4], YTrain)
```

We will use $k = 11$ to fit a model to our training dataset.

The prediction can be obtained using the predict method of our model:

```
y_pred = knnclf.predict(XTest[:, 2:4])
```

The predictions of our model can be obtained with the predict method.

In Figure 6.3 we show a contour map of the three Iris classes obtained using KNN based on sepal width and petal length.

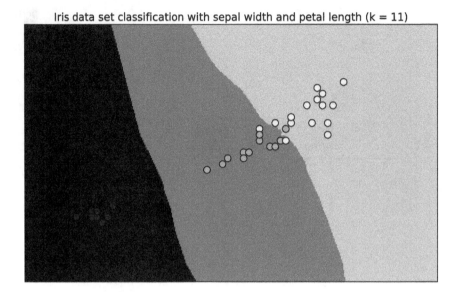
Iris data set classification with sepal width and petal length (k = 11)

The data points correspond to the flowers in the testing set, coloured according to their class. We can see some instances where the flowers have been misclassified.

Figure 6.3: KNN classification of the Iris dataset based on sepal width and petal length for $k = 11$. The shading areas correspond to the classification mapping obtained by the algorithm. We can see some misclassifications in the upper right-hand corner of the plot.

Let us construct a confusion matrix for our classifier:

```
> from sklearn.metrics import confusion_matrix

> confusion_matrix(YTest,y_pred)

array([[12,  0,  0],
       [ 0, 14,  2],
       [ 0,  2, 15]])
```

A confusion matrix can be obtained with the confusion_matrix function from the module metrics.

As we can see four instances for classes 2 and 3 (two each) have been misclassified. We can obtain a classification report as follows:

```
> from sklearn.metrics \
    import classification_report
> print(classification_report(YTest, y_pred))
      precision    recall  f1-score   support
   0       1.00      1.00      1.00        12
   1       0.88      0.88      0.88        16
   2       0.88      0.88      0.88        17

avg / total    0.91    0.91    0.91        45
```

The classification_report function provides information about the precision and recall of the classification model trained.

The report gives us information about the precision and recall for our classifier as well as the F_1 score which provides a measure of accuracy based on the precision and recall as follows:

$$F_1 = 2\frac{(precision)(recall)}{(precision + recall)},$$ (6.6)

and ranges between 1 (best) and 0 (worst).

6.3   Classification with Logistic Regression

WE ARE FAMILIAR WITH THE concept of regression as discussed in Chapter 4. We saw that regression is a supervised machine learning task that enables us to obtain predictions for continuous variables. In contrast, logistic

Regression is used for continuous variables. For logistic regression think of categorical variables.

regression is used in the prediction of a discrete outcome and therefore best suited for classification purposes. Logistic regression is in effect another generalised linear model that uses the same basic background as linear regression. However, instead of a continuous dependent variable, the model is regressing for the probability of a (binary) categorical outcome. We can then use these probabilities to obtain class labels for our data observations.

Logistic regression predicts the probability of a categorical outcome.

Let us recall that a linear regression model is the conditional mean of the outcome variable $\mathbf{Y}$ given the value of the covariate $\mathbf{X}$ noted as $E(\mathbf{Y}|\mathbf{X})$ and is given by:

$$E(\mathbf{Y}|\mathbf{X}) = \beta \mathbf{X}, \tag{6.7}$$

For a univariate case $\beta \mathbf{X} = \beta_0 + \beta_1 x$.

where we assume that this conditional mean is a linear function taking values between $-\infty$ and ∞.

In logistic regression, we are interested in determining the probability that an observation belongs to a category (or not) and therefore the conditional mean of the outcome variable must lie in the interval $[0, 1]$. We need to extend the linear regression model to map the outcome variable into that unit interval. We can do this with the help of the sigmoid function defined as:

We need to map our outcomes to values between 0 and 1.

$$g(z) = \frac{e^z}{1 + e^z}. \tag{6.8}$$

The logistic function.

We show a plot of this function in Figure 6.4 where we can see that the domain of the function is $(-\infty, \infty)$ and its range is $[0, 1]$, as required. This function is also known as the **logistic function**.

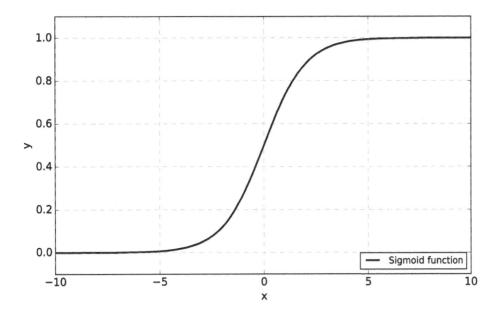

Our transformed regression model (for a univariate case) is
therefore given by:

$$E(\mathbf{Y}|\mathbf{X}) = P(\mathbf{X}) = \frac{\exp{(\boldsymbol{\beta}\mathbf{x})}}{1+\exp{(\boldsymbol{\beta}\mathbf{x})}}. \qquad (6.9)$$

Our transformed model.

It is also relevant to mention an important transformation
related to the logistic function, namely the **logit function**
or **log-odds function**. The logit of a probability given by
$p \in [0,1]$ is:

$$\text{logit}(p) = \ln\left(\frac{p}{1-p}\right). \qquad (6.10)$$

The logit or log-odds function.

The interpretation of the quantity above is such that if p
represents a probability, the logit of the probability is the

logarithm of the odds. Conveniently, the difference between the logits of two probabilities is the logarithm of the odds ratio, which can be expressed in terms of simple addition and substraction.

The difference between the logits of two probabilities is the logarithm of the odds ratio.

We can show that we can recover our linear model by applying the logit function to Equation (6.9):

$$
\begin{aligned}
\ln\left(\frac{P(\mathbf{X})}{1-P(\mathbf{X})}\right) &= \ln P(\mathbf{X}) - \ln\left(1 - P(\mathbf{X})\right) \\
&= \ln\left(\frac{e^{\beta x}}{1+e^{\beta x}}\right) - \ln\left(1 - \frac{e^{\beta x}}{1+e^{\beta x}}\right) \\
&= \ln\left(e^{\beta \mathbf{X}}\right) - \ln\left(1 + e^{\beta \mathbf{X}}\right) + \ln\left(1 + e^{\beta \mathbf{X}}\right) \\
&= \beta \mathbf{X}.
\end{aligned}
\tag{6.11}
$$

The logit function can therefore be a very useful tool in interpreting the results obtained from the logistic regression algorithm.

Another important difference between the linear and the logistic regressions is the error term. In the case of the linear regression the error term follows an independent normal distribution with zero mean and constant variance.

Normal distributions are also called Gaussian

In logistic regression, however, the outcome variable can take only two values: Either 0 or 1. This means that instead of following a Gaussian distribution it follows a Bernoulli one. The Bernoulli distribution corresponds to a random variable that takes the value 1 with probability p and 0 with probability $q = 1 - p$.

Think of a coin tossing experiment.

Our logistic regression model given by Equation (6.9) will provide us the probability that an observation belongs to a binary class or not. We need a threshold that helps us carry out the classification task, and a typical example is as follows:

$$y_i = \begin{cases} 1, & \text{if} \quad P(\mathbf{X}) \geq 0.5. \\ 0, & \text{otherwise.} \end{cases} \qquad (6.12)$$

These are typical thresholds applied alongside a logistic regression.

Since the logistic regression model is a generalised linear model for outcomes in $(0,1)$ the solution can be implemented using Ordinary Least Squares and still use measures such as R^2. In cases where the outcome is exactly 0 or 1 other methods, such as maximum likelihood estimation, are needed.

See Chapter 4 for more information.

Let $\mathbf{x}$ be a set of training data and $\mathbf{y}$ correspond to the labels for the dataset. We are interested in maximising the log-likelihood of the data and so for i samples we have that the coefficients are given by:

$$\beta \leftarrow \max_{\beta} \sum_i \ln P(\mathbf{X}_i, \beta). \qquad (6.13)$$

The objective function for logistic regression.

We can also apply regularisation to the logistic regression model and thus the optimisation problem can be written as:

$$\beta \leftarrow \max_{\beta} \sum_i \ln P(\mathbf{X}_i, \beta) - \lambda ||\beta||_n, \qquad (6.14)$$

The objective function for regularised logistic regression.

where $|| \cdot ||_n$ is the L-n norm. In that way we can achieve the penalisation for high coefficients as discussed for the linear regression in Section 4.9.

6.3.1   Logistic Regression Interpretation

AS WE HAVE MENTIONED ABOVE, the interpretation of the results from logistic regression is better understood in terms of the odds and the odds ratio. We know that the odds of an event with probability p is given by:

$$odds = \frac{p}{1-p},$$

(6.15)

The odds of an event with probability p.

which is effectively the ratio of the probability that the event will take place (p) to the probability that it will not $(1-p)$.

In a univariate case, from Equation (6.11) we have that:

$$\frac{P(\mathbf{x})}{1-P(\mathbf{x})} = \exp\left(\beta_0 + \beta_1 x\right).$$

(6.16)

The odds for the univariate case.

Remember that in linear regression, we interpret the parameter β_1 as the change in the target variable given a unit change in the covariate x. Let us see how we can use this to interpret our coefficients in the logistic regression. We consider a univariate case and a change of one unit in the variable x. The odds would be given by

Let us consider a unit change in x.

$$odds(x+1) = \exp\left(\beta_0 + \beta_1(x+1)\right).$$

(6.17)

Let us now take the odds ratio of Equations (6.16) and (6.17):

$$OR = \frac{\exp\left(\beta_0 + \beta_1 x\right)}{\exp\left(\beta_0 + \beta_1(x+1)\right)} = e^{\beta_1}. \qquad (6.18)$$

We then calculate the odds ratio.

If we now take the logarithm of the odds ratio we have that it is equal to the coefficient β_1:

$$\ln(OR) = \beta_1, \qquad (6.19)$$

and this means that the coefficients obtained from logistic regression can be interpreted as the change in the logit function for a unit change in the covariate. In other words, the odds ratio of a binary event gives the increase in likelihood of an outcome if the event occurs. Similarly, the odds ratio in logistic regression represents how the odds change with a unit increase in a given feature, holding all other features constant.

The coefficients correspond to the change in the logit function for a unit change in x.

For example, imagine that you estimate a logistic regression model with features x_1 and x_2 such that the model can be expressed as $10.0145 + 0.25x_1 + 0.04x_2$. The effect of the odds of a unit increase in x_1 is given by $\exp(0.25)=1.284$, and this means that the odds increase by about 28%, regardless of the value of x_2.

An example of the interpretation of the logistic regression coefficients.

Before we move on to see an example of the application of logistic regression let us address an important point regarding the number of classes in our problem. Our discussion has concentrated on a problem with two classes (A and B, 1 and 0, etc). Logistic regression can be used in a multiclass setting too and a typical strategy is the so-called *one-versus-the-rest* strategy. In this case the modelling is done

It is possible to use logistic regression in cases involving more than two classes.

by using one classifier per class so that for each of our classifiers, the chosen class is modelled against the rest of the classes. In other words, if we have a problem with three classes 0, 1 and 2, we will need three classifiers:

1. Classifier one for class 0 versus 1 and 2

2. Classifier two for class 1 versus 0 and 1

3. Classifier three for class 2 versus 0 and 1

6.3.2   Logistic Regression in Action

WE ARE GOING TO USE a breast cancer dataset that contains cases from a study at the University of Wisconsin Hospital by W. H. Wolberg and O. L. Mangasarian[5]. The dataset is available at the UCI Machine Learning Repository under "Breast Cancer Wisconsin (Original) Dataset" and is available at `https://archive.ics.uci.edu/ml/datasets/Breast+Cancer+Wisconsin+(Original)`[6].

We are interested in classifying breast tissue samples into benign or malignant cases. The dataset contains 699 instances with an ID column and 10 features as follows:

1. Sample code number (ID number)

2. Clump Thickness (1- 10)

3. Uniformity of Cell Size (1- 10)

4. Uniformity of Cell Shape (1- 10)

5. Marginal Adhesion (1- 10)

[5] Mangasarian, O. L. and W. H. Wolberg (1990, Sep.). Cancer diagnosis via linear programming. *SIAM News* 25(5), 1 & 18

[6] Lichman, M. (2013b). UCI Machine Learning Repository, Wisconsin Breast Cancer Database. `https://archive.ics.uci.edu/ml/datasets/Breast+Cancer+Wisconsin+(Original)`. University of California, Irvine, School of Information and Computer Sciences

6. Single Epithelial Cell Size (1- 10)

7. Bare Nuclei (1- 10)

8. Bland Chromatin (1- 10)

9. Normal Nucleoli (1- 10)

10. Mitoses (1- 10)

11. Class (2 for benign, 4 for malignant)

We assume that the data has been pre-processed into a CSV file with appropriate names for the columns detailed above. We can then load it into Python as follows:

```
bc = pd.read_csv(u'./breast-cancer-wisconsin.csv')
bc = bc.dropna()
```

The dropna method drops instances with missing information.

We have dropped any instances where no information is available, leaving us with 683 instances; we can see how they are distributed with the help of describe:

```
> bc['Class'] = bc['Class'].astype('category')
> bc['Class'].describe()

count      683
unique       2
top          2
freq       444
Name: Class, dtype: int64
```

We cast the column Class as a categorical variable.

As we can see, the most frequent class is '2' (benign) with 444 instances. Let us prepare our data by separating the labels from the rest of the dataset:

```
X = bc.drop(['Class'], axis=1)

X = X.values

Y_raw = bc['Class'].values
```

We separate the labels from the rest of the dataset by dropping the appropriate column.

It would make our task easier to use '0' and '1' as the labels for our classes instead of the labels used by in the original dataset. This can easily be done with LabelEncoder as follows:

```
from sklearn import preprocessing

label_enc = preprocessing.LabelEncoder()

label_enc.fit(Y_raw)

Y = label_enc.transform(Y_raw)
```

The LabelEncoder method lets us re-assign the encodings used for our classes.

It is possible to see the classes with label_enc.classes_ and very importantly we can invert the encoding with label_enc.inverse_transform().

Let us split our data into training and testing as we have been doing so far:

```
import sklearn.model_selection as cv


XTrain, XTest, YTrain, YTest =\

ms.train_test_split(X, Y,\

test_size=0.3, random_state=1)
```

Scikit-learn implements logistic regression in the `linear_model` module and is called `LogisticRegression`. We are going to use regularisation in our model and we can choose between $L1$ and $L2$ penalties. The hyperparameter in this case is implemented as `C` and it corresponds to the inverse of the regularisation strength. This means that the smaller the value of `C`, the stronger the penalty. We will use `GridSearchCV` to determine the best values for these two parameters:

The logistic regression model is implemented in the `linear_model` module. The hyperparameter is denoted by `C`.

```
from sklearn.linear_model \
import LogisticRegression


pen_val = ['l1','l2']
C_val = 2. ** np.arange(-5, 10, step=2)
grid_s = [{'C': C_val, 'penalty': pen_val}]
model = LogisticRegression()


from sklearn.model_selection\
import GridSearchCV


cv_logr = GridSearchCV(estimator=model,\
param_grid=grid_s,\
cv=ms.KFold(n_splits=10))
```

As with other examples we have seen, we can search for optimal values of the hyperparameter with `GridSearchCV`.

The fitting can now be done as follows:

```
cv_logr.fit(XTrain, YTrain)
best_c = cv_logr.best_params_['C']
best_penalty = cv_logr.best_params_['penalty']
```

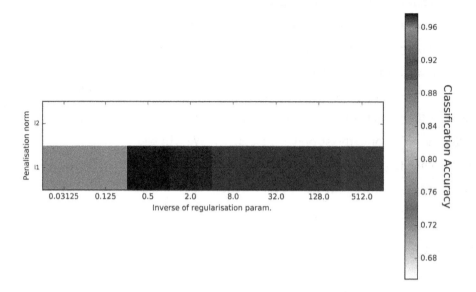

Figure 6.5: A heatmap of mean cross-validation scores for the Logistic Regression classification of the Wisconsin Breast Cancer dataset for different values of C with L1 and L2 penalties.

In this case the grid search was carried out with 10 folds, over the hyperparameter values defined, as well as *L*1 and *L*2 penalties returns the following parameters:

```
> print(''The best parameters are:\
cost={0} and penalty={1}''.\
format(best_c, best_penalty))


The best parameters are: cost=0.5 and penalty=l1
```

Figure 6.5 shows a heatmap of the parameter search we have performed. It is now possible for us to use the parameters above to create a classifier for our problem and train it to be used on unseen data. All we have to do is create an instance of the logistic regression model using the parameters obtained:

```
b_clf = LogisticRegression(C=best_c,\
penalty=best_penalty)


b_clf.fit(XTrain, YTrain)
```

We are instantiating a model with the parameters found above.

The classifier can now be used to predict the class in our testing dataset with the `predict` method. We obtain the probabilities assigned to each instance with the `predict_proba` method:

```
predict = b_clf.predict(XTest)


y_proba = b_clf.predict_proba(XTest)
```

`predict` gives us the predicted class, where as `predict_proba` gives us the probabilities.

Please note the `y_proba` is an array with two columns, each column indicates the probability that the instance belongs to either of the two classes.

Let us take a look at the classification accuracy score:

```
> print(b_clf.score(XTest, YTest))


0.960975609756
```

The accuracy of the model can be seen with the `score` method.

The coefficients for the model can be obtained with the `.coef_` method:

```
> print(b_clf.coef_)


[[ -2.76001862e-06   3.34121074e-01
    3.64527317e-01   3.44557750e-01
    2.99474883e-02  -1.24661155e-01
    3.49887361e-01   1.30369406e-01
    2.94834342e-01   1.56018289e-01]]
```

The coefficients are stored in the coef_ method.

Remember that the odds ratios are obtained by taking the exponential of the coefficients. They tell us how a unit increase or decrease in a variable affects the odds of having a malignant mass:

```
> print(np.exp(b_clf.coef_))

[[ 0.99999724  1.39671224
   1.43983326  1.41136561
   1.03040042  0.88279598
   1.41890772  1.13924915
   1.34290388  1.16884758]]
```

Taking the exponential of the coefficients lets us interpret the results.

For instance we can expect the odds of having a malignant mass to increase by about 43% if the measure for "Uniformity of Cell Shape" increases by one unit.

Finally, let us take a look at obtaining the ROC and AUC measures. In the case of ROC we can use roc_curve whose inputs are the true labels of the instances and the target scores such as probability estimates of the positive class

The next step is to calculate the ROC and AUC.

or confidence values. Let us use the y_proba estimates we
calculated above:

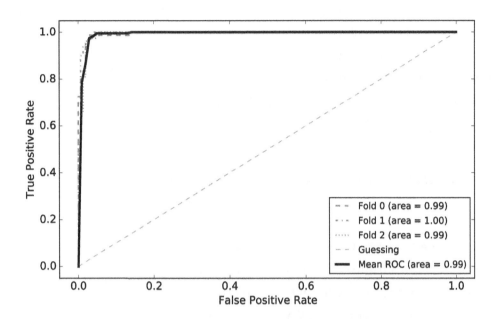

Figure 6.6: ROC curves obtained
by cross-validation with $k = 3$
on the Wisconsin Breast Cancer
dataset.

```
from sklearn.metrics import roc_curve, auc


fpr, tpr, threshold=roc_curve(YTest, y_proba[:,1])


plt.plot(fpr, tpr)
```

and the AUC can be calculated with the true positive and
false positive rates obtained above:

```
> print(auc(fpr, tpr))

0.992167919799
```

In Figure 6.6 we show ROC curves obtained by using cross-validation on the dataset using 3 folds for simplicity. We also show the mean ROC calculated from the cross-validation and compare it with the random classifier given by the diagonal line shown.

It is possible to obtain various ROC plots with the aid of cross-validation.

6.4   Classification with Naïve Bayes

PROBABILITY IS AT THE CENTRE of many everyday life applications: From weather to sports, passing by finance and science. As a theory, we owe much to the French mathematicians Pierre de Fermat and Blaise Pascal who used to discuss problems related to games of chance with each other[7]. As such, in the early seventeenth century, it was widely believed that it was not possible to calculate (predict) the outcome of rolling a dice for example. We now know that is not really the case.

[7] Devlin, K. (2010). *The Unfinished Game: Pascal, Fermat, and the Seventeenth-Century Letter That Made the World Modern*. Basic ideas. Basic Books

In the previous section we used logistic regression to estimate the probability that a data instance belongs to a particular class and, based on that, decide on the label that should be assigned to that instance. We assumed that we knew what was meant by probability, i.e. a number between 0 and 1 that indicates how likely it is for an event A to occur. We denote the probability of event A as $P(A)$.

We denote the probability of and event A as $P(A)$.

From a traditional standpoint, probability is presented in terms of a **frequentist** view where data instances are drawn from a repeatable random sample with parameters that remain constant during the repeatable process. These assumptions enable us to determine a frequency with which an event occurs. In contrast, the **Bayesian** view takes the approach that data instances are observed from a realised sample, and that the parameters are unknown. Since the repeatability of the data sample does not hold under this view, the Bayesian probability is not given in terms of frequencies, but instead it represents a state of knowledge or a state of "belief".

There are two probability schools of thought: *frequentist* and *Bayesian*.

In the Bayesian view, probability reflects a state of knowledge.

The Bayesian approach is named after the 18th century English statistician and Presbyterian minister Thomas Bayes who is known for formulating a theorem that bears his name: *Bayes' theorem*. Bayes' stardom in the scientific world is grounded on a posthumous publication[8] presented to the Royal Society by his friend Richard Price.

[8] Bayes, T. (1763). An essay towards solving a problem in the doctrine of chances. *Philosophical Transactions 53*, 370–418

Bayes' theorem states that the probability of a particular hypothesis is given by both current information (data) and prior knowledge. The prior information may be the outcome of earlier experiments or trials, or even educated guesses drawn from experience. This is the reason why many frequentist practitioners have shunned the Bayesian approach for centuries. Nonetheless, Bayesian statistics has stood the test of time[9] and demonstrated its usefulness in many applications. Let us consider the set of all possible events, Ω, which is our *sample space*. Event A is a member of the sample space, as is every other event. The probability of

[9] McGrayne, S. (2011). *The Theory that Would Not Die: How Bayes' Rule Cracked the Enigma Code, Hunted Down Russian Submarines, & Emerged Triumphant from Two Centuries of Controversy*. Yale University Press

the sample space is $P(\Omega) = 1$. The probability $P(A)$ is given by:

$$P(A) = \frac{|A|}{|\Omega|},\qquad(6.20)$$

where $|A|$ denotes the cardinality of A. We show a Venn diagram of this situation in Figure 6.7.a). If $|A|$ were to have equal cardinality to Ω then the probability of A would be at most 1.

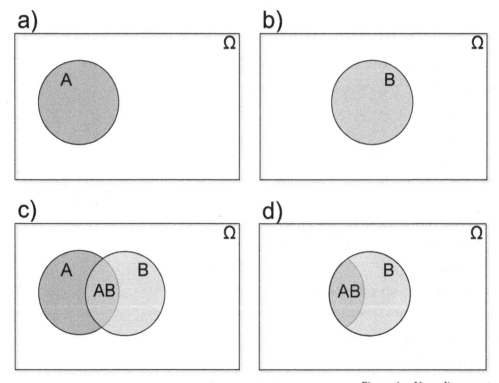

Figure 6.7: Venn diagrams to visualise Bayes' theorem.

For a different event B, we have a similar situation (shown in Figure 6.7.b) and the probability $P(B)$ is:

$$P(B) = \frac{|B|}{|\Omega|}.$$ (6.21)

We are using the frequentist view in this case.

Thinking about the dataset we analysed in the previous section, let us consider A to be the set of women with breast cancer, and B the set of women whose test for cancer was positive. Any woman in either of the sets above would be interested to know if their test was positive *and* if they actually have cancer. This would be given by the intersection of A and B, namely $A \cap B$, or AB for short. This would be the situation depicted in Figure 6.7.c). So, how do we calculate that? Actually, in the same fashion as before:

The probability of events A and B happening together is related to the intersection $A \cap B$.

$$P(AB) = \frac{|AB|}{|\Omega|}.$$ (6.22)

We are using AB as a shorthand for $A \cap B$.

Notice that the calculation above aims at excluding the following events: 1) women with cancer and with a negative test result ($A - AB$), and 2) women with a positive result but without cancer ($B - AB$).

All in all, for this example, the question of real importance is whether for a randomly selected woman, *given that the test is positive, what is the probability that she has cancer?* In terms of our Venn diagrams (see Figure 6.7.d)) the question above is equivalent to *given that we are in region B, what is the probability that we in fact are in region AB?*.

The next question to ask is the probability of event A, given event B, i.e. $P(A|B)$.

We denote the conditional probability of A given B as $P(A|B)$, and it can be calculated as follows:

$$P(A|B) = \frac{|AB|}{|B|} = \frac{\frac{|AB|}{|\Omega|}}{\frac{|B|}{|\Omega|}} = \frac{P(AB)}{P(B)}. \qquad (6.23)$$

It is also possible to calculate $P(B|A)$:

$$P(B|A) = \frac{P(BA)}{P(A)}, \qquad (6.24)$$

and for our example it will give us the probability of a randomly selected woman to test positively given that she has the condition. Incidentally, if the events A and B are independent from each other, information about one does not affect the other and thus $P(A|B) = P(A)$. This means that $P(AB) = P(A)P(B)$.

> If events A and B are independent from each other, information from one does not affect the other.

The numerators of Equations (6.23) and (6.24) are actually the same. This means that we can write the following: two as $P(A|B)P(B) = P(B|A)P(A)$, which in turn can be rearranged as

> $P(AB) = P(BA)$.

$$P(A|B) = \frac{P(B|A)P(A)}{P(B)}. \qquad (6.25)$$

> This is Bayes' theorem.

This result is what we know as **Bayes' theorem**. We call $P(A|B)$ the *posterior* probability, $P(A)$ the *prior* probability and $P(B|A)$ the likelihood. Bayes' theorem can be thought of as a rule that enables us to update our belief about a hypothesis A in light of new evidence B, and so our posterior belief $P(A|B)$ is updated by multiplying our prior

> $P(A|B)$ is the posterior probability and $P(A)$ is the prior.

belief $P(A)$ by the likelihood $P(B|A)$ that B will occur if A is actually true. This rule has had a number of very successful applications and perhaps one of my favourite ones is a very early one by Pierre Simon Laplace to determine the mass of Saturn[10].

[10] Laplace, P. and A. Dale (2012). *Pierre-Simon Laplace Philosophical Essay on Probabilities: Translated from the fifth French edition of 1825 With Notes by the Translator*. Sources in the History of Mathematics and Physical Sciences. Springer New York

In Bayesian statistics we tend to work along the following steps:

1. We first set out a probability model for an unknown parameter of interest, and include any prior knowledge about the parameter, if available at all

2. Using the conditional probability of the parameter on observed data, we update our knowledge of this parameter

3. We then evaluate the fitness of the model to the data and check the conclusions reached based on our assumptions

These are some typical steps followed in Bayesian statistics.

4. We may then decide to start over again using the new information gained as our starting prior

The priors we use in the first step above reflect the best approximation to what we are interested in modelling. This information may come from expert information, researcher intuition, previous studies, other data sources, etc. In the past, Bayesian analysis would require a lot of manual computation. These days, the help of computers has lowered this barrier considerably.

The priors reflect the best approximation to what we are interested in modelling.

6.4.1   Naïve Bayes Classifier

LET US NOW SEE HOW Bayes' theorem can be used in a
classification task. A typical example, and one that most
of us are familiar with, is the task of determining if a new
email that arrives to our inbox is a good email ("ham")
or a bad email ("spam"). Let us imagine that we have an
existing corpus of email data that can be decomposed
in such a way that we end up with a dataset containing
$\{x_i\} = x_1, x_2, \ldots, x_n$ features and we are interested in
classifying new emails as belonging to the class S of spam
emails.

A collection of text documents is
called a *corpus*.

We can use Bayes' theorem to calculate the conditional
probability of an email belonging to class S given the
observation of features $\{x_i\}$ as:

$$P(S|\{x_i\}) = \frac{P(\{x_i\}|S)\,P(S)}{P(\{x_i\})}. \qquad (6.26)$$

We are calculating the probability
of an email to be spam given the
features we observe in the corpus.

The value of the likelihood $P(\{x_i\}|S)\,P(S)$ can be obtained
directly from the training data: We know that we have spam
emails correctly labelled as such. We can therefore obtain
the probability of observing the set of features $\{x_i\}$ given
the spam email sample we have at our disposal. Similarly,
the value of the prior $P(S)$ can also be obtained from the
data as the training dataset includes both spam and ham.
Finally, $P(\{x_i\})$ is a constant that does not depend on
the class S and can be incorporated at the end of all our
computations.

The majority of the quantities
needed can be readily calculated.

On the whole, the most difficult part in the calculation outlined above is the estimation of the likelihood function $P(x_1, x_2, \ldots, x_n|S)$. To get this value we need to have a lot of data in order to cover every possible combination of features $\{x_i\}$ and obtain a good estimation. This may be achievable, but remember that a model is just a reasonable approximation to reality. As such, we can make a suitable assumption to simplify the computation of this quantity.

The hardest part is the likelihood function $P(\{x_i\}|S)$.

Naïvely, we can assume for example that the features $\{x_i\}$ are conditionally independent from each other. If that is the case we can re-write the likelihood function as:

$$P(x_1, x_2, \ldots, x_n|S) \simeq P(x_1|S) \cdot P(x_2|S) \cdot \ldots \cdot P(x_n|S). \quad (6.27)$$

We can make the calculation tractable by assuming that the features are conditionally independent from each other.

It is this naïve assumption that makes the calculation above tractable and gives the name to this classification algorithm.

6.4.2   Naïve Bayes in Action

THE EXAMPLE WE ARE GOING to follow for the naïve Bayes classifier is based on text classification. Instead of a spam detector, we will work with Twitter data coming from my own streams @quantum_tunnel, and @dt_science. Our task is to determine if a given tweet is about "data science" or not. The data is available[11] at https://dx.doi.org/10.6084/m9.figshare.2062551.v1.

You can follow me on Twitter: @quantum_tunnel and @dt_science.

[11] Rogel-Salazar, J. (2016a, Jan). Data Science Tweets. 10.6084/m9.figshare.2062551.v1

The corpus to be used in this task is already split into training and testing datasets. The training data contains 324 labelled tweets related to "data science" in a file called

Train_QuantumTunnel_Tweets.csv with three columns: A
"data science" label, the date when the tweet was published
and the actual published text of the tweet. The testing
dataset is not labelled and contains 163 tweets. Let us load
the training data into a Pandas dataframe:

The corpus of tweets is already split into training and testing sets.

```
import pandas as pd


train = pd.\
read_csv('Train_QuantumTunnel_Tweets.csv',\
encoding='utf-8')
```

In Python 3 it is important to specify the encoding of a text file. In this case UTF-8.

We can inspect the data by slicing the dataframe:

```
print(train[62:64])
```

Tweets 62 and 63 actually read:

> Tweet 62: *And that is Chapter 3 of "Data Science and Analytics with Python" done... and moving on to the rest! Super chuffed! #BookWriting'*

The content of the tweets has not been pre-processed. We will clean the input in the next steps.

> Tweet 63: *See sklearn trees with #D3 https://t.co/UYsioXbcbu*

We would like to pre-process the text of the tweets to get rid
of URLs and hashtags. We can do so by writing a function
as follows:

```
import re
def tw_preprocess(tw):
    ptw = re.sub(r''http\S+'', '''', tw)
    ptw = re.sub(r''#'', '''', ptw)
    return ptw
```

We will use the regular expressions package re.

We can now apply the function to the relevant column in the Pandas dataframe:

```
train['Tweet'] = train['Tweet'].\
apply(tw_preprocess)
```

The apply method of a Pandas dataframe allows us to pre-process our corpus.

We are interested in characterising the tweets by the words that appear in each of them, and so these words will be our features. To that end, we need to tokenise our tweets and generate a term-document matrix. In other words, a matrix whose rows correspond to documents (in this case tweets) and its columns correspond to terms (words). We can get this done with the help of CountVectorizer in Scikit-learn:

A term-document matrix is a matrix whose rows correspond to documents and its columns to words.

```
from sklearn.feature_extraction.text \
import CountVectorizer

vectoriser = CountVectorizer(lowercase=True,\
stop_words='english',\
binary=True)
```

CountVectorizer enables us to create our term-document matrix.

We can now apply our vectoriser to the training tweets as follows:

```
X_train = vectoriser.\
fit_transform(train['Tweet'])
```

We use our vectoriser to transform our corpus.

The result is a large sparse matrix, so printing it is not a good idea. Nonetheless, we can still see the vocabulary that has been gathered from the training set by fit_transform. This can be done with get_feature_names:

```
> vectoriser.get_feature_names()[1005:1011]

['putting', 'python',

'quantitative', 'quantum',

'quantum_tunnel', 'question']
```

We can list the vocabulary created with the help of get_feature_names.

We are now in a position to create our model using the sparse matrix generated by our vectoriser and the labels provided with the training dataset:

```
from sklearn import naive_bayes

model = naive_bayes.MultinomialNB().\
fit(X_train, list(train['Data_Science']))
```

We are using the MultinomialNB algorithm from naive_bayes.

We have not employed cross validation as yet, but we can certainly take a look at the scores obtained with cross validation as follows:

```
> import sklearn.model_selection as ms
> ms.cross_val_score(naive_bayes.\
MultinomialNB(), X_train, train['Data_Science'],\
cv=3)

array([ 0.74311927,  0.77777778,  0.72897196])
```

In this case we use cross-validation with 3 sets, enabling us to see the scores attained.

We can see the confusion matrix that we obtain for the training dataset with the model we just created:

```
> from sklearn.metrics import confusion_matrix
> confusion_matrix(train['Data_Science'],\
model.predict(X_train))


array([[195,    1],
       [  0, 128]])
```

We discussed confusion matrices in Section 6.1.1.

Finally, we can now apply our model to the testing dataset provided. We need to load the data and let us apply the same preprocessing we used for the training dataset:

```
test = pd.\
read_csv('Test_QuantumTunnel_Tweets.csv',\
encoding='utf-8')
test['Tweet'] = test['Tweet'].\
apply(tw_preprocess)
```

We need to load the training dataset and apply the same transformation used for training.

We now have to apply the vectoriser to the testing dataset via the `transform` method and to finally apply the `predict` method of the model:

```
X_test = vectoriser.transform(test['Tweet'])
pred = model.predict(X_test)
```

We have to use the `transform` method of our vectoriser before running our prediction.

We can also see the probability assigned to each tweet as follows:

```
print(pred)
pred_probs = model.predict_proba(X_test)[:,1]
```

The `predict_proba` method lets us see the probabilities assigned to each of the testing tweets.

Let us check for example the probability assigned to the tweet with id 103:

```
> pred_probs[102]


0.99961933738874065
```

Remember that Python starts counting from 0.

Let us see the text of this tweet:

> *Finished writing Chapter 4 for my DataScience and analytics with Python book. Moving on to discussing some classification analysis*

The rest of the data can be checked in a similar fashion. Notice that we have used a very small corpus for this demo, but you can see how powerful Bayes' theorem is, even when making the naïve assumption we described earlier on in this chapter. Furthermore, in this example we were not particularly careful when cleaning the data, for instance we left numbers and punctuation in the corpus, also we did not apply any stemming on the text either.

More sophisticated pre-processing of the corpus is left to the reader as an exercise: Get rid of punctuation, numbers, etc.

6.5   Summary

CLASSIFICATION IS ANOTHER VERY IMPORTANT tool to have under our jackalope data scientist belt. Together with clustering, these techniques let us understand our data and obtain insights which in turn can be used to get actionable predictions. We saw how classification is a supervised learning task that exploits information about the classes contained in our data in order to make predictions about unseen data instances.

We discussed three well-known techniques, namely: k-Nearest-Neighbours (or KNN), Logistic Regression and Naïve Bayes. In KNN we use the similarity of a given data instance and its closest data points. Logistic regression extends the linear model we discussed in Chapter 4 using a sigmoid function to estimate the probability that a given data instance belongs to a class. Finally, naïve Bayes uses Bayes' theorem to update the posterior probability that a data instance belongs to a class, given that certain features are observed in the data.

Finally, we saw how confusion matrices, true and false positives and negatives, as well as the recall and fallout measures let us evaluate how well our classifiers are performing. The same applies to the Receiver Operator Characteristic (ROC) curve and the area under this curve (AUC). We saw the algorithms described above applied to a variety of problems, from cancer to twitter data as well as wine and flower datasets.

7

Decisions, Decisions: Hierarchical Clustering, Decision Trees and Ensemble Techniques

THE USE OF DIAGRAMS AND illustrations in science and business is nothing new. Along with tables and inscriptions, they provide us with useful representations of concepts and data we need to communicate. They make it easier for us to organise our knowledge and information. A resource that has been widely used for centuries and across cultural and disciplinary landscapes is that of the tree[1]. It is easy to see the appeal: Trees enable us to represent information in a hierarchical manner that is easy to follow.

[1] Lima, M. and B. Shneiderman (2014). *The Book of Trees: Visualizing Branches of Knowledge.* Princeton Architectural Press

In this chapter we will address a few techniques that, in one way or another, have been inspired by the metaphorical representation of a tree: From roots to branches and leaves. First we will talk about a clustering algorithm that organises groups of data in a hierarchical manner. Then we will talk about decision trees, a tool that is widely used in decision analysis applications and operations research. Towards

The techniques we will discuss in this chapter have been inspired by trees.

the end of the chapter we will combine trees in ensembles giving rise to random forests.

7.1   *Hierarchical Clustering*

THE WORD HIERARCHY EVOKES THE idea of a system where information is ranked according to a relative status, giving rise to different levels. Hierarchical clustering is an unsupervised learning task whose goal is to build a hierarchy of data groups. The hierarchy can be built from the "bottom-up". In this case each data instance starts in its own cluster, and we successively merge these clusters as we go up in the levels of the hierarchy. This is known as **agglomerative clustering**. In contrast, **divisive clustering** takes the opposite approach, starting with all data instances and performing splits as we go down the levels of the hierarchy. The results of hierarchical clustering are presented in a tree-like structure called a **dendrogram**.

Hierarchical clustering is an unsupervised learning task.

The result of hierarchical clustering are represented by a dendrogram.

As we know, clustering relies on the existence of a similarity measure among data instances. In a dendrogram, data points are joined together from the most similar (closest) to the most different (further apart). Let us recall our discussion from Section 5.2 about the k-means algorithm. We need to determine k different clusters with three things: The number of clusters k, an initial assignation of data points to the clusters, and a distance or similarity measure $d(x_i, x_j)$. In hierarchical clustering we only need one thing: A similarity measure among groups of data points.

In a dendrogram, data points are joined together from the most similar to the most different.

Let us take a look at the agglomerative clustering approach, where our starting point is a situation where each of our N data instances is on their own cluster. Given the similarity measure $d(x_i, x_j)$, we iteratively merge the two closest groups, repeating until all data instances end up in a single cluster. The point at which two clusters are joined up is called a *node*. At each of the nodes of the resulting tree we are effectively segmenting our data in a sequential manner.

In hierarchical clustering we only need to define a similarity measure.

A node is a point where two clusters are joined.

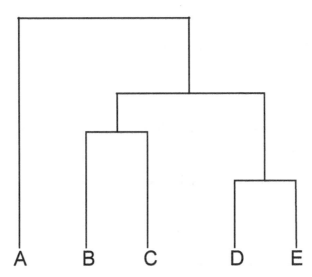

A   B   C   D   E

Figure 7.1: A dendrogram is a tree-like structure that enables us to visualise the clusters obtained with hierarchical clustering. The height of the clades or branches tells us how similar the clusters are.

It is easy to see that the similarity among merged groupings decreases monotonically at every node created by agglomerative clustering. The more dissimilar two clusters are, the more distant they are. This is actually depicted in a dendrogram: The height of each branch, or *clade*, tells us how similar or different the groupings are from each other, the greater the height, the greater their dissimilarity. We show a sample dendrogram in Figure 7.1. We see that

The height of the branches tells us how similar the clusters are.

clusters D and E are more similar to each other than B and C. Also, cluster A is significantly different from the rest of the other clusters.

At the start of the agglomerative process, finding similar data points is straightforward as we are comparing individual data points. However, once we have more than one instance per cluster we need to define what we mean by similarity among groups or clusters. Given two clusters F and G, we have a few choices:

As we know, defining what we mean by similarity is a very important step in the process.

1. **Single linkage**: We define the similarity by the closest instance pair: $d_{SL}(F,G) = \min d(x_i, x_j)$ with $x_i \in F$ and $x_j \in G$

2. **Complete linkage**: The similarity is given by the farthest pair: $d_{CL}(F,G) = \max d(x_i, x_j)$ with $x_i \in F$ and $x_j \in G$

Some similarity measures used in hierarchical clustering.

3. **Group average**: The similarity is given by the mean similarity between F and G:
$$d_{GA}(F,G) = \frac{1}{|F||G|} \sum_i \sum_j d(x_i, x_j) \text{ with } x_i \in F \text{ and } x_j \in G$$

Be aware that single linkage may lead to a situation where a sequence of close data points gives rise to a chain. This is because clusters tend to merge very early on in the process. By the same token, complete linkage may not merge close groups in cases where there are far-apart outliers in the clusters. Finally, group average depends on the scale of the similarities taken into account, and over all is a good compromise between the three choices detailed above.

As usual, there are advantages and disadvantages with each of these choices.

Another alternative for cluster merging is the so-called Ward method based on an analysis of variance (ANOVA) approach. In this method, at each stage two clusters would merge when they provide the smallest increase in the combined error sum of squares from one-way univariate ANOVAs that may be performed for each feature.

An alternative is the use of the Ward method.

Finally, remember that hierarchical clustering is an unsupervised algorithm and in cases where groupings do not actually exist in the data, the algorithm will still impose a hierarchy. Cross validation may provide some clues as to whether the clusters we obtain may actually make sense.

Remember that hierarchical clustering is an unsupervised task.

7.1.1 Hierarchical Clustering in Action

LET US USE THE IRIS dataset to demonstrate the use of hierarchical clustering with Python. Although Scikit-learn provides hierarchical clustering implementations such as AgglomertiveClustering in this case we are going to use the hierarchy module in SciPy.

We will use SciPy with the Iris dataset.

Let us retrieve the Iris dataset and load the data to a suitable variable:

```
from sklearn.datasets import load_iris
iris = load_iris()


X = iris.data
```

By now, we are very well acquainted with loading the Iris dataset.

From SciPy, we will be using the linkage and dendrogram functions from hierarchy, and from distance we will call the function pdist:

```
from scipy.cluster.hierarchy import linkage
from scipy.cluster.hierarchy import dendrogram
from scipy.spatial.distance import pdist
```

We will use the functions linkage, dendrogram and pdist from SciPy.

The `dpist` function lets us calculate the pairwise distances between the features in our data, whereas `linkage` is used to run hierarchical clustering on a distance matrix:

```
X_dist = pdist(X)


X_link = linkage(X, method='ward')
```

We calculate pairwise distances among our data points and run the hierarchical clustering algorithm.

We are using the Ward method in our agglomerative clustering task.

We can obtain a measure of how the pairwise distances in our data compare to those implied by the hierarchical clustering. This is can be done with the *cophenetic coefficient*[2]. The better the clustering preserves the original distances, then the closer this coefficient is to 1:

[2] Farris, J. S. (1969). On the cophenetic correlation coefficient. *Systematic Biology 18*(3), 279–285

```
> from scipy.cluster.hierarchy import cophenet
> coph_cor, coph_dist = cophenet(X_link, X_dist)
> print(coph_cor)


0.872601525064
```

A cophenetic coefficient closer to 1 indicates that the clustering preserves the original pairwise distances of the data points.

It is sometimes important to know what clusters have been merged at each iteration. This information is actually contained in the matrix returned by the `linkage` function. The *i*-th entry of this matrix tells us what clusters have

The linkage function provides information about what clusters have been merged together.

been merged at iteration i (first two entries), as well as their distance (third entry) and the number of samples contained (fourth entry). For example, we can see what data points were merged at the first iteration as follows:

```
> print(X_link[0])

[  9.,   34.,    0.,    2.]
```

The 9th and 34th data instances were merged first.

which indicates that the data instances with indices 9 and 34 were merged at the first iteration.

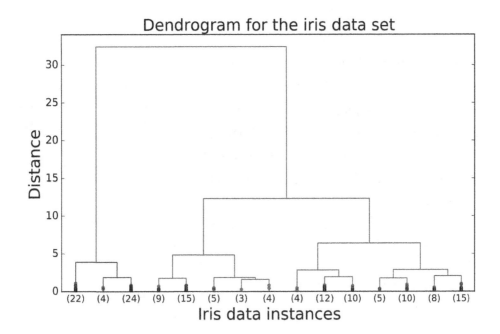

We are showing the dendrogram generated for the Iris data set in Figure 7.2. The numbers in brackets along the x-axis

Figure 7.2: Dendrogram generated by applying hierarchical clustering to the Iris dataset. We can see how three clusters can be determined from the dendrogram by cutting at an appropriate distance.

correspond to the number of data instances that are under each clade. Remember that we are starting from a point where each data instance is a cluster (i.e. the final leaves in our tree). In this case we have a total of 150 entries, and showing all those branches may not be that practical. The dendrogram function provides us with a way to truncate the diagram. In this case we are showing the last $p = 15$ merged clusters:

```
import matplotlib.pyplot as plt
dendrogram(X_link, truncate_mode = 'lastp',\
p=15, show_contracted = True)
plt.show()
```

The dendrogram function lets us plot the result of our hierarchical clustering.

Finally, we can obtain the labels that the hierarchical clustering generates. This can be done with fcluster, which takes a threshold corresponding in this case to the distance where we want to truncate our dendrogram. From Figure 7.2 we can see that at a height lower than 10 we have three distinct clusters. Let us then pick a threshold of 9 and use that with in fcluster:

The fcluster function lets us truncate the dendrogram and assign labels to our data points.

```
from scipy.cluster.hierarchy import fcluster
max_d = 9
clusters = fcluster(X_link, max_d,\
criterion='distance')
```

The array clusters contains the labels 1, 2, or 3 assigned to our data points.

An alternative way of choosing the threshold can be done with the *inconsistency* measure. It compares the height h of each cluster merge to the average, and normalises the difference to the standard deviation:

```
from scipy.cluster.hierarchy import inconsistent
depth = 6
incons_measure = inconsistent(X_link, depth)
```

An alternative way to choose the threshold is with the use of the inconsistency measure.

The matrix returned by `inconsistent` contains the average, standard deviation, count and inconsistency for each merge. We can use this measure to obtain our clusters. However, this method has a high reliance on having choosing the right parameters for threshold and depth:

```
clusters_incons = fcluster(X_link, t=8,\
criterion='inconsistent', depth=15)
```

The inconsistency measure can also be used to obtain our labels.

7.2 *Decision Trees*

CONTINUING WITH THE INSPIRATION PROVIDED by trees, in this section we are going to explore one of the most well-known classification algorithms, and one that is actually named after its botanical counterpart. A decision tree is an algorithm described as non-parametrical because it does not require us to make any assumptions about parameters or distributions before starting our classification task. It is also a hierarchical technique: The model is built in such a way that a sequence of ordered decisions about the values of the data features results in assigning a class label to any given data instance.

Decision trees are among the most well-known classification algorithms.

Think of these decisions as questions regarding the data.

You may be asking yourself how to recognise a decision tree, even from a very long way away. Well, you will be happy to know that a decision tree is probably better known by its

You will surely recognise a tree, even from a very long way away.

diagrammatic representation, where rules guide us through the chart to decide a final outcome. The tree structure consists of nodes and edges, where the nodes represent the test conditions or questions we need to consider for classifying our data. The edges represent the answers or outcomes to the questions we are asking.

A decision tree is a type of a directed acyclic graph.

A good tree must have a solid root, and a decision tree is no exception. We have a **root node** which is the node that has no incoming edges, and two or more outgoing edges. An **internal node** has one incoming edge and two or more outgoing ones: Internal nodes represent test conditions at every given level. Finally a **leaf node** has one incoming edge an no outgoing ones; leaf nodes correspond to class labels, as there are no further outcomes.

Even a larch has a root.

And as such, our decision trees have roots, as well as branches and even leaves.

In Chapter 2 we created a Pandas dataframe for some animals, we detailed the number of limbs and dietary habits in Table 2.4. Let us expand the information by adding a column that labels the data. We show this in Table 7.1 and we will use this information to construct our first decision tree.

Let us revisit our first Pandas dataframe from Chapter 2.

Animal	Limbs	Herbivore	Class
Python	0	No	Reptile
Iberian Lynx	4	No	Mammal
Giant Panda	4	Yes	Mammal
Field Mouse	4	Yes	Mammal
Octopus	8	No	Mollusc

Table 7.1: Dietary habits and number of limbs for some animals.

It is possible to evaluate all possible combinations of the test conditions for a given dataset and construct all possible decision trees from there. For example, we can start by considering as our root node the test of whether or not the animal is a herbivore. We can continue by asking how many limbs the animal has, and so on. The result is the decision tree shown in Figure 7.3. Note that we could have started by asking first if the animal has 4 limbs or not. This would have resulted in having immediately a leaf node where all the mammals in our dataset get classified in one go.

We start here by asking whether the animal in question is a herbivore or not. Then we ask about the number of limbs, and so on.

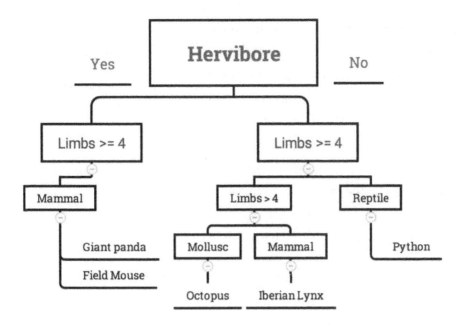

Figure 7.3: A simple decision tree built with information from Table 7.1.

It stands to reason that constructing all possible decision trees is not a very practical thing to do. Instead, we are interested in applying a greedy algorithm that lets us find a local optimum to solve our classification problem in a much faster way.

One possibility is the application of Hunt's algorithm[3], which grows a decision tree in a top-down fashion by recursively splitting the dataset into smaller and smaller data subsets. The aim is to minimise the cost of the classification task: On the one hand taking into account the misclassification of a data instance, and on the other one determining the value of a feature exhibited by the data instance.

[3] Hunt, E. B., J. Marin, and P. J. Stone (1966). *Experiments in induction.* New York: Academic Press

Selecting smaller subsets of data is an easy thing to do. However, we require smaller subsets that actually lead to leaf nodes that let us classify data instances accurately. With that in mind, we therefore need a measure to partition our dataset at every node. That measure is called *purity*: A partition is said to be 100% pure if all of the the data instances in the subset belong to the same class and this means that it is not necessary to split the data any further. This would have been the case in our example above, using Table 7.1, if we had started splitting the data by the number of limbs: animals described in that table as having 4 limbs are all mammals.

Not only is it necessary to select smaller subsets, but also they need to lead to a more accurate classification.

A partition is 100% pure if all its data points belong to the same class.

For a binary classification problem with classes A and B, and a dataset with N_n data instances at node n, Hunt's algorithm proceeds as follows:

1. If node n is pure, in other words, all its data instances N_n belong to class A, we are done as we effectively have a leaf node for class A

2. If node n is not pure, we need to split the dataset further. We need to create a test condition that enables a partition

Hunt's algorithm in a few simple steps.

in the dataset. This means that the node becomes an internal node

3. We run the test condition and assign each of the instances N_n to one of the two child nodes created from node n

4. These steps are applied recursively to each and every child node

Step 2 above indicates that we need to create a test condition. We prefer a condition that gives us the best split in terms of purity for each of the child nodes. In other words, the larger the purity the better our classification. We need to compare the impurity of the parent node with the impurity of each of the child nodes generated after splitting our data.

The test condition we need should give us the best split in terms of purity.

With a total of c classes, let us denote the fraction of records that belong to class i at a given node n as $p(i|n)$. Some popular measures of impurity include the following:

- Entropy:

$$H(n) = - \sum_{i=1}^{c} p(i|n) \log_2 p(i|n). \qquad (7.1)$$

A typical measure used is the entropy.

The entropy of a pure node is zero as $\log_2(1) = 0$. It reaches is maximum when all classes have equal proportions and in the particular case of a binary classifier the maximum is 1.

- Gini impurity:

$$G(n) = 1 - \sum_{i=1}^{c} p^2(i|n). \qquad (7.2)$$

The Gini impurity is also widely used.

The Gini impurity of a pure node is zero. As it is the case for the entropy measure, the Gini impurity has a maximum value of 1 when all classes have equal proportions. The values of the Gini impurity range between 0 and 1 irrespective of the number of classes involved.

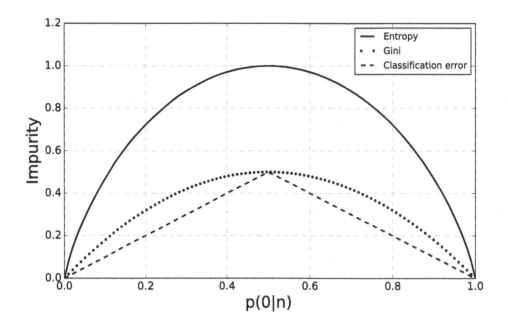

Figure 7.4: A comparison of impurity measures we can use for a binary classification problem.

- Classification error:

$$E(n) = 1 - \max_i p(i|n). \qquad (7.3)$$

Classification error.

Once again, the classification error for a pure node is zero. Its values range between 0 and 1.

Figure 7.4 shows the behaviour of the impurity measures described above.

With any of the impurity functions $I(n)$ mentioned above, we are now in a position to compare the purity of the parent and children nodes with the **gain**:

$$\Delta = I(\text{parent}) - \sum_{\text{children}} \frac{N_j}{N} I(j\text{-th child}). \qquad (7.4)$$

The gain enables us to compare the purity of the parent and children nodes.

When the impurity measure employed is the entropy, the gain is actually called the **information gain**.

Decision tree algorithms are not exempt from overfitting and the worst case scenario would be one where we end up with a number of leaves equal to the number of data instances. We can restrict our tree growth to binary splits exclusively (CART algorithm) or by imposing a penalty on the number of outcomes (C4.5 algorithm). A popular regularisation employed is the *gain ratio*, which takes into account the entropy of the distribution of data instances into the children nodes:

Decision trees are not exempt from overfitting.

$$GR = \frac{\Delta_{info}}{-\sum \frac{N_j}{N} \log_2 \left(\frac{N_j}{N} \right)}. \qquad (7.5)$$

The gain ratio is a popular regularisation technique employed with decision trees.

The gain ratio reduces the bias towards multi-valued features as it considers the number and size of nodes when choosing a feature for splitting our data.

Overfitting can also be avoided by determining criteria that stop us from continuing splitting of our data. A trivial example is the case when all the data instances in a node belong to the same class. We can set a minimum threshold on the gain and stop when the information on a branch becomes unreliable as we are no longer achieving a gain

Pre-pruning our tree consists on setting a minimum threshold on the gain.

above the threshold imposed. In this case we are *pre-pruning* our tree and although it prevents overfitting, its calibration is not straightforward and may also stop growth prematurely.

An alternative is to work on a fully grown tree and apply *post-pruning*. We effectively examine the tree bottom-up and simplify subtrees either by replacing them with a single node (subtree replacement) or with a simpler subtree (subtree raising).

Post-pruning our tree is another alternative.

7.2.1   Decision Trees in Action

THE TITANIC AND ITS FATEFUL maiden voyage from Southampton, England to New York City, US is well-known to all of us. Famous even before sailing, the vessel was effectively a floating city, and Her passengers came from all backgrounds in society: From the wealthy elite to hopeful immigrants looking to start afresh in America. On the night of April 14th, 1912 the Titanic struck an iceberg, ultimately sinking the vessel. The Titanic sent distress signals by telegraph and, although there were other ships nearby, no assistance was promptly provided.

We are all familiar with the fateful maiden voyage of the Titanic.

Using information about passengers of the Titanic, we are interested in building a model based on a decision tree to say something about the chances of surviving the disaster. The data we are going to use can be obtained from the Kaggle competition *"Titanic: Machine learning from Disaster"*[4] at the following URL: https://www.kaggle.com/c/titanic.

[4] Kaggle (2012). Titanic: Machine Learning from Disaster. https://www.kaggle.com/c/titanic

The training data provided contains 891 records with the following attributes:

- Survived: 0= No; 1 = Yes

- Pclass: Passenger Class (1 = 1st; 2 = 2nd; 3 = 3rd)

- Name: Passenger name

- Sex: (female; male)

- Age: Passenger age

- SibSp: Number of Siblings/Spouses Aboard

- Parch: Number of Parents/Children Aboard

- Ticket:Ticket Number

- Fare: Passenger Fare

- Cabin: Cabin

- Embarked: Port of Embarkation (C = Cherbourg; Q = Queenstown; S = Southampton)

The attributes in the Titanic dataset.

Let us load the data into a Pandas dataframe and get rid of some information that is incomplete. In this case, we will ignore the Ticket and Cabin columns, and drop instances without values in the rest of the dataframe:

There is some missing data in the dataset.

```
titanic = pd.read_csv(u'./Data/train.csv')


titanic = titanic.drop(['Ticket','Cabin'], axis=1)


titanic = titanic.dropna()
```

In this case we are dropping the Ticket and Cabin columns, as well as records with missing data.

These transformations leave us with 712 data instances to work with.

There is some extra information that may help us navigate the use of the various attributes in the dataset. Famously, the ship did not have enough lifeboats for all the crew and passengers. Furthermore, the lifeboats used were not used at full capacity. Passengers in the upper classes were the first to be helped out, leaving the rest to fend for themselves.

No pun intended...

Let us perform some basic exploratory analysis.

We can assume that the old adage of *"Women and children first"* may have applied. With that in mind, let us perform some data exploration and see the percentage of passengers who survived the disaster, categorised by travelling class and gender:

```
Pclas_pct =\
pd.crosstab(titanic.Pclass.astype('category'),\
titanic.Survived.astype('category'),\
margins=True)


Pclas_pct['Percent'] =\
Pclas_pct[1]/(Pclas_pct[0]+Pclas_pct[1])


Sex_pct =\
pd.crosstab(titanic.Sex.astype('category'),\
titanic.Survived.astype('category'),\
margins=True)


Sex_pct['Percent'] = \
Sex_pct[1]/(Sex_pct[0]+Sex_pct[1])
```

We first construct a cross-tabulation of the travel class v survival with `crosstab`.

We can use Pandas to calculate the percentage of survivors per class.

What about the percentage of survivors per gender?

Let us now look at the numbers:

```
> print(Pclas_pct['Percent'], Sex_pct['Percent'])


Pclass
1.0    0.652174
2.0    0.479769
3.0    0.239437
All    0.404494
Name: Percent, dtype: float64
Sex
female    0.752896
male      0.205298
All       0.404494
Name: Percent, dtype: float64
```

We can now take a look at the percentages calculated above.

We can see from the data provided that about 65% of passengers travelling in 1st class, compared to only 23% of those traveling in 3rd class. In terms of gender, 75% of female passangers survived, compared to only 20% of males. In other words, you had a better chance of surviving if you were a female passenger travelling in 1st class accommodation.

A more thorough data exploration analysis can be done. Go ahead, give it a go!

For our modelling we will use a subset of the features provided and concentrate on three things: Class, gender and age. Remember that Scikit-learn only accepts numerical values as data, and in this case the Sex feature is given as text, i.e "female" and "male". Let us pre-process our data with Pandas to obtain dummy variables to encode the information in numerical labels:

Remember that Pandas only takes numerical values. We need to encode categorical values appropriately.

```
titanic = pd.concat([titanic,\
pd.get_dummies(titanic['Sex'])], axis=1)
```

The get_dummies function lets us do this very easily.

This appends two columns to our dataframe, one called female and the other one called male, indicating the values with 0 or 1.

We are now ready to start our modelling. Scikit-learn provides a decision tree model in DecisionTreeClassifier, accepting various parameters such as the impurity criterion taking values such as entropy and gini. We can also affect the pruning with parameters such as max_depth and min_samples_leaf, the former determines the the maximum depth (levels) of the tree, whereas the latter determines the minimum number of data instances required to split an internal node.

We are making use of the DecisionTreeClassifier algorithm implemented in Scikit-learn.

Let us use GridSearchCV to determine the best values of the maximum depth and minimum number of samples. We start by splitting our data into training and testing:

```
X = titanic[['Pclass','Age','female']]
Y = titanic['Survived']


import sklearn.model_selection as ms


XTrain, XTest, YTrain, YTest =\
ms.train_test_split(X, Y,\
test_size= 0.3, random_state=1)
```

Let us split our data into training and testing. Notice that we are only including the column female as a 0 here implies a 1 in the male column.

We can now define the values over which our search will be performed, and instantiate our decision tree model to use entropy as the impurity measure:

```
depth_val = np.arange(2,11)
leaf_val = np.arange(1,31, step=9)


from sklearn import tree


grid_s = [{'max_depth': depth_val,\
'min_samples_leaf': leaf_val}]


model = tree.DecisionTreeClassifier(criterion=\
'entropy')
```

We will perform our parameter search over the depth of the tree and the threshold of data for node splitting.

Finally, we instantiate our model using entropy for the impurity criterion.

It is possible now to run our search and use the best parameters found:

```
from sklearn.model_selection import GridSearchCV


cv_tree = GridSearchCV(estimator=model,\
param_grid=grid_s,\
cv=ms.KFold(n_splits=10))


cv_tree.fit(XTrain, YTrain)


best_depth = cv_tree.best_params_['max_depth']


best_min_samples = cv_tree.\
best_params_['min_samples_leaf']
```

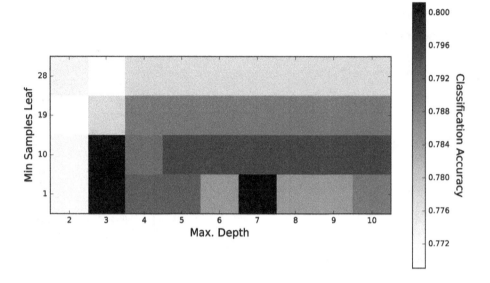

Figure 7.5: Heatmap of mean cross-validation scores for the decision tree classification of the Titanic passengers for different values of maximum depth and minimum sample leaf.

In this case the parameters turned out to be 3 for the maximum depth and 1 for the minimum number of data instances. We show a heatmap of the classification score in Figure 7.5. Let us now apply our model to the testing subset we created above:

```
model = tree.DecisionTreeClassifier(\
criterion='entropy',\
max_depth=best_depth,\
min_samples_leaf=best_min_samples)


TitanicTree = model.fit(XTrain, YTrain)
survive_pred = TitanicTree.predict(XTest)
survive_proba = TitanicTree.predict_proba(XTest)
```

This is the moment of truth, we will apply our model to the testing subset and see how it performs.

The confusion matrix we obtain from our model above is as follows:

```
> from sklearn import metrics
> metrics.confusion_matrix(YTest, survive_pred)


array([[104,  22],
       [ 23,  65]])
```

We can take a look at the confusion matrix for this classifier.

The score obtained is:

```
> print(TitanicTree.score(XTest, YTest))


0.789719626168
```

As well as the overall score on the testing dataset.

Scikit-learn enables us to generate a Graphviz visualisation of the tree we have generated with the export_graphviz method in tree. We show our tree in Figure 7.6.

Graphviz is an open source graph visualisation software. You need a separate install to visualise the tree.

```
tree.export_graphviz(TitanicTree,\
out_file='TitanicTree.dot',\
max_depth=3, feature_names=X.columns,\
class_names=['Dead','Survived'])
```

This code will generate a .dot file that can be visualised with Graphviz.

Finally, we can apply our model to the hold-out data provided by the test dataset from the Kaggle competition. We need to preprocess our data first as we did for the training dataset:

In this case we are using the Kaggle dataset as a holdout. Our model has not seen this dataset at all yet.

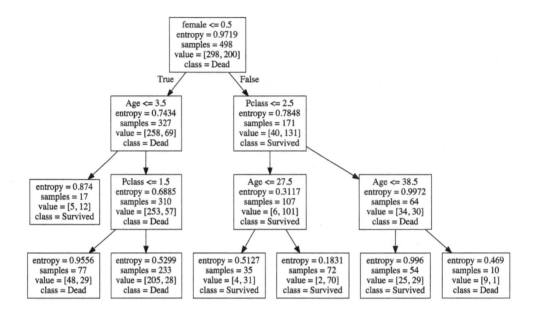

Figure 7.6: Decision tree for the Titanic passengers dataset.

```
titanic_test = pd.read_csv(u'./Data/test.csv')

titanic_test = titanic_test.drop(['Ticket',\
'Cabin'], axis=1)

titanic_test =titanic_test.dropna()

titanic_test = pd.concat([titanic_test,\
pd.get_dummies(titanic_test['Sex'])], axis=1)
```

We need to apply the same transformations we performed on the training set.

Now we finish with obtaining our predictions:

```
X_holdout = titanic_test[['Pclass','Age',\
'female']]


survive_holdout = TitanicTree.predict(X_holdout)
```

Finally, we run our model on the holdout dataset.

7.3   Ensemble Techniques

MANY OF US MAY HAVE encountered in our travels the enticing competition of guessing the number of jelly beans, candy or coins in a jar. It is an easy one to grasp: All you have to do is provide your estimate of the number, leave your details and, if you get it right, you are awarded a prize together with a lifetime of newsletters to your inbox. This sort of competition has been running for quite a long time, but not many of us stop to consider the statistical implications of the estimates provided. Not unless you are Francis Galton.

Yes, the same Francis Galton of regression fame we discussed in Chapter 4.

In a county fair held at Plymouth, England, Galton encountered a version of the competition above, but instead of jelly beans, the contestants had to guess the weight of a fat ox[5]. He reports that the average guess of $1,207$ lbs among 800 contestants is actually rather close to the actual weight of said ox ($1,198$ lbs). He remarked that the middlemost estimate expresses the *vox populi*. These days we refer to it as the "wisdom of crowds".

[5] Galton, F. (1907). Vox populi. *Nature* 75(1949), 450–451

One single estimate of the weight of the ox (or the number of coins in the jar) may not be accurate or even close to the real value. However, when taking into account a diversity of independent opinions and aggregating them, the estimate becomes more accurate. Of course the wisdom of crowds is not infallible but it is worth considering. In fact we do this when seeking a second or even third opinion on medical diagnostics. Furthermore, citizen science projects such as

Simply take a look at panicked investors in a bursting bubble market.

Zooniverse[6] and Fold-it[7] have demonstrated that the collaboration of the general public and nonprofessional amateurs provides an invaluable contribution to the scientific endeavour.

[6] Zooniverse. Projects. `https://www.zooniverse.org/projects`
[7] Fold-it. Solve puzzles for science. `https://fold.it/portal/`

A similar effect has been successfully shown when combining classifiers in machine learning. We call these combinations **ensemble techniques**. In that respect, an ensemble is a set of individually trained base classifiers, such as decision trees, whose predictions are combined to determine the class of unseen data. As in the example of the ox investigated by Galton, the results obtained from an ensemble are generally more accurate than those of the individual classifiers.

An ensemble is a set of individually trained classifiers whose predictions are combined.

The base classifiers can be any algorithm that provides us with a result that is slightly better than random guessing and hence is referred to as a **weak learner**. For a binary classifier, a weak learner will provide us with a function that correctly classifies a record with at least a probability $\frac{1}{2} + \epsilon$, with a small and positive value for ϵ. In contrast, a **strong learner** provides us with a very successful algorithm to classify our records.

Classifiers can be categorised into weak and strong learners.

Our task is to find a way to aggregate the results provided by our weak learners and obtain a strong one. Nonetheless, in order for the ensemble classifier to be strong, and therefore outperform a weak base classifier, we need the base classifiers be accurate and they must show diversity in their misclassifications, in other words, their classification errors must occur on different training records. The

We aim to combine weak learners and get a strong one.

Base classifiers must outperform random guessing.

accuracy requirement indicates the low bias of the model, whereas diversity suggests that the weak learners are uncorrelated. For these reasons, a decision tree is the epitome of a base classifier: The flexibility provided by the number of labels that we can take into account when growing our tree, for example, enables us to generate various base classifiers.

A decision tree is a classic example of a weak learner.

Let us consider the advantages that are provided by an ensemble classifier in terms of a supervised learning task. We are interested in making accurate predictions of the true class of new data by learning the classifier h. If our training data is small, the base classifier on its own will find it difficult to converge to h. An ensemble classifier can help by "averaging out" the predictions of the individual base classifiers, improving convergence.

An ensemble classifier can help by averaging out the predictions of the individual base learners.

There are also advantages in terms of computation: An exhaustive search over all possible classifiers is a hard problem. In this case, even with enough data at our disposal, finding the best h is difficult. The use of an ensemble made out of several base classifiers with different starting conditions can help with our search.

This is why we used a heuristic in Section 7.2 to grow our decision trees.

Finally, a single base classifier may provide a classification function that does not adapt very well to the actual h. A good example is the canonical decision tree, whose decision boundaries are given by rectilinear partions of the feature space, as shown in Figure 7.7.a). In this case, if the true boundary is given by a diagonal line, we would need our base classifier to provide an infinitely large number of

A single base classifier may not provide a good approximation to the actual decision boundary. Instead, an ensemble can provide a better approximation.

a)

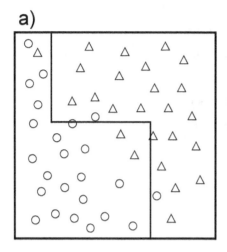

b)

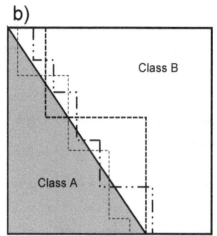

Figure 7.7: Decision boundaries provided by a) a single decision tree, and b) by several decision trees. The combination of the boundaries in b) can provide a better approximation to the true diagonal boundary.

segments. However, the ensemble classifier can provide a good approximation to the boundary, as can be seen from Figure 7.7.b).

The idea of aggregating base classifiers is quite straight forward. Given an initial training dataset, we need to build a set of base classifiers. Each of these n classifiers need to be trained with appropriate data. A way to do this is shown diagrammatically in Figure 7.8, where the initial training data is used to create multiple datasets according to a given sampling distribution. This distribution determines how likely we are to select a particular record for training. Each of these datasets is used to train weak learners to predict the class of unseen data, and finally the models are aggregated.

Our training data is used to create multiple datasets to train the base classifiers.

The ensemble classifier can be constructed by applying different mechanisms. For instance, we can manipulate a) the training set, b) the class labels, or c) the learning algorithm. We shall talk about these mechanisms in more

There are different ways to construct our ensemble.

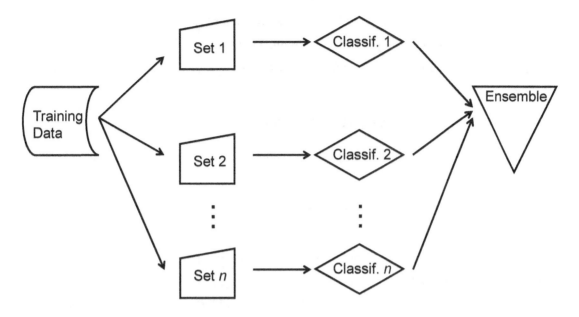

Figure 7.8: A diagrammatic view of the idea of constructing an ensemble classifier.

detail. First, let us get a clearer picture of how ensemble methods achieve better results than a single base classifier.

Let us consider a binary classification task with labels A and B. We have total of 10 new data records whose true label is A for all of them. We build three classifiers whose accuracy is 60%. We obtain the predicted labels returned by each of the three classifiers and take a majority vote of the labels to construct our ensemble. An instance of results is given in Table 7.2.

Classifier	Predicted Class									
1	A	B	A	A	B	B	A	B	A	A
2	B	A	A	A	A	A	B	A	B	B
3	A	B	A	B	A	B	A	B	A	A
Ensemble	A	B	A	A	A	B	A	B	A	A

Table 7.2: Predicted classes of three hypothetical binary base classifiers and the ensemble generated by majority voting.

In the particular example we show in Table 7.2, the majority vote has boosted the accuracy to 70%. All in all, for a majority vote with the three classifiers above we expect the following outcomes:

- All three classifiers are wrong: $(1 - 0.6)^3 = 0.4^3 = 0.064$

- Two classifiers are wrong: $3(0.6)(0.4)^2 = 0.288$

- Two classifiers are correct $3(0.6)^2(0.4) = 0.432$

- All three classifiers are correct: $0.6^3 = 0.216$

Expected outcomes of the ensemble considered in Table 7.2.

As we can see, in about 43% of the cases the majority vote provides the correct label. From the figures above, the ensemble will be correct about 64% of the time (43% + 21%), performing better than the 60% of a single base classifier. Using a larger number of classifiers will provide a better classification.

Using a larger number of base classifiers provide a better result.

We have mentioned above that correlation among the classifiers has an impact on the ensemble. Let us show some empirical examples: If all of our three classifiers have correlated outcomes we may see no improvement. Consider the outcomes shown in Table 7.3 where classifiers 2 and 3 have an accuracy of 70% and are highly correlated in their predictions. The ensemble shows no real improvement when taking the majority vote (70%).

Classifiers with correlated outcomes may show no improvement.

Let us now compare the result above with that of three classifiers that have very different performances and with uncorrelated results as shown in Table 7.4. As we can see, the ensemble with a majority vote achieves in this case an accuracy of 90%.

In contrast, uncorrelated outcomes in the classifiers may show an improved accuracy.

Classifier	Predicted Class									
1 (Accuracy 80%)	A	A	A	A	A	B	A	B	A	A
2 (Accuracy 70%)	B	A	A	A	A	A	B	A	A	B
3 (Accuracy 70%)	B	A	A	A	A	A	B	A	A	B
Ensemble	B	A	A	A	A	A	B	A	A	B

Table 7.3: Predicted classes of three hypothetical binary base classifiers with high correlation in their predictions.

Classifier	Predicted Class									
1 (Accuracy 80%)	A	A	A	A	A	B	A	B	A	A
2 (Accuracy 60%)	A	B	A	B	B	A	B	A	A	A
3 (Accuracy 70%)	B	A	A	A	A	A	B	A	A	B
Ensemble	A	A	A	A	A	A	B	A	A	A

Table 7.4: Predicted classes of three hypothetical binary base classifiers with low correlation in their predictions.

In the examples above we have used a simple majority voting to decide the outcome for the ensemble. In the following sections we discuss some of the different ways in which we can construct our ensemble classifier.

7.3.1 Bagging

BAGGING STANDS FOR BOOTSTRAP AGGREGATION and it is an ensemble technique that involves the manipulation of the training dataset by resampling. Given an initial training dataset with N records, in bagging we create multiple training datasets of size N, by sampling uniformly with replacement. This means that some records may be picked more than one time during the process, and some may not even appear at all.

Bagging stands for bootstrap aggregation.

Note that in bagging some records may appear more than once in the engineered datasets.

We build classifiers on each bootstrap sample and take a
majority vote across the classifiers. Since some of the newly
created datasets may contain repeated instances of the data,
as well as missing some records altogether, we end up with
a situation where some classifiers will have an error rate
higher than a classifier that uses all the raw data.

Similarly, some data points may be missing altogether, giving rise to a collection of weak learners.

Nonetheless, as we have seen in the example shown in
Table 7.4, when combining these classifiers we may have a
result with an accuracy that is better than that of a single
classifier on its own. Breiman has shown[8] that bagging is
an effective methodology on learning algorithms where
small changes in the training set have large impacts in the
resulting predictions. These algorithms are called "unstable"
and some typical examples include decision trees or neural
networks.

[8] Breiman, L. (1996). Bagging predictors. *Machine Learning* 24(2), 123–140

The increased accuracy from bagging comes from the
reduction in the variance of the individual classifier,
improving therefore our generalisation error. If the
classifiers are stable, the error incurred comes mainly from
their bias and bagging may not be effective. Since we are
resampling the data with replacement, bagging does not
focus on particular data instances of the training data. This
means that we are less prone to overfitting.

The increased accuracy from bagging comes from a reduction in the variance of the single classifier.

7.3.2   Boosting

FROM THE METHODOLOGY USED IN bagging we now know
that specific data records are not given any preferential
treatment. However, it is worth considering changes that

we can implement if we wanted to focus our attention on specific particular records in the training dataset. A possibility is to change the sampling distribution of the training records. Boosting can be described as an iterative methodology that adapts the sampling of the data in order to concentrate on records that have been misclassified in previous iterations.

Boosting is an iterative method that adapts data sampling to focus on misclassified records.

In the initial iteration we start with a uniform distribution assigning equal weights to our N records. At the end of the first round we change the weights to emphasise those records that were misclassified. In that way, boosting produces a series of classifiers whose inputs are chosen based on the performance of the previous family of classifiers in the series. The final prediction of the series is computed by a weighted vote, depending on the training errors of the individual base classifiers.

We start in the same way we did for bagging.

Boosting produces a series of classifiers with inputs based on the outcome of the previous family of classifiers.

Boosting is aimed at building, with each iteration, classifiers that are better at predicting the label of a data instance than the classifiers in the previous iteration. The sampling is done with replacement and we may encounter situations were some particular records do not show in a given training subset. This is not a problem as these overlooked records are more likely to be misclassified and as a result they will be given a higher weight in later iterations, forcing the ensemble to correct for these mistakes. We can see that on each iteration, the base classifiers will concentrate on records that are harder and harder to classify, facing more difficult learning tasks as the task progresses.

We use sampling with replacement.

Overlooked records are more likely to be misclassified. This is corrected in later iterations as the weight of these instances will be higher.

A popular implementation of boosting is the Ada-Boost algorithm[9], or adaptive boosting. It uses either the approach of selecting a set of data points based on the probability of the instances themselves, or by using all the data instances and weighting the error of each instance by its own probability. Ada-Boost is a fast algorithm with no parameters to tune, except for the number of iterations. Nonetheless, if the base classifiers are complex, Ada-Boost may lead to overfitting. Remember that Ada-Boost may be susceptible to noise.

[9] Freund, Y. and R. Schapire (1997). A decision-theoretic generalization of on-line learning and an application to boosting. *J. Comp. and Sys. Sciences* 55(1), 119–139

7.3.3   Random Forests

A SINGLE TREE PROVIDES A good shade, but the canopy of a group of trees is difficult to beat. If we grow a group of various decision trees as our base classifiers, and enable their growth to use a random effect we end up with a *random forest*[10]. We have seen that bagging produces better results than a single base classifier. If the base classifier is for instance a decision tree, the various trees we generate with bagging have different predictions because they use different training sets.

Difficult to beat a group of trees for shade.

[10] Breiman, L. (2001). Random forests. *Machine Learning* 45(1), 5–32

If we were to introduce changes in the way our trees are grown, for example by randomly selecting not only the data instances that are included in each subset, but also by randomly selecting the features to use, we can get base classifiers that are uncorrelated with each other. In effect, random forests add a layer of randomness to bagging with decision trees: On top of growing each tree using each

Not only do we randomly select data instances, but also the features used to grow our trees.

bootstrap sample, in a random forest each node of the tree is split among a subset of randomly chosen features. The result is a classifier that performs very well and is robust to overfitting. A random forest algorithm is very straightforward to understand as it only has two parameters: 1) the number of features in the random subset at each node, and 2) the number of trees in the forest. The simplest forest we can grow consists of selecting at each node a small number of features to perform the split. We then simply allow the trees to grow to maximum size without pruning. This methodology is known as *Forest-RI* or Random Input selection.

In cases where there is a small number of features, it is possible to create new features by taking random linear combinations of the existing ones. These new features are then used to grow the trees. This process is called *Forest-RC* or Random Combinations. If instead we have a large number of predictors we have to bear in mind that the available feature set will be very different from one node to the next.

We can go one step forward in our inclusion of randomness into the growth of our trees and instead of simply picking the best split among the chosen features at a node, the thresholds for the split are selected at random for each chosen feature. The best of these randomly-generated thresholds becomes the rule for splitting. These type of random forests are known as *extremely randomised trees*[11]. As with any of the other ensemble techniques we have outlined, the greater the correlation among the trees we grow, the

The result is a robust classifier.

We have two parameters in a random forest:

- Number of features at each node
- Number of trees in the forest

We can grow the trees by random input selection or by random feature combinations.

[11] Geurts, P., D. Ernst, and L. Wehenkel (2006). Extremely randomized trees. *Machine Learning 63*, 3–42

greater the random forest error rate. We would therefore prefer trees that are uncorrelated among themselves. Running a random forest algorithm is fast, having the advantage of being robust against unbalanced or even missing data. However, with datasets that are particularly noisy, they tend to overfit.

Remember that we prefer uncorrelated base classifiers.

7.3.4   Stacking and Blending

APART FROM THE ENSEMBLE TECHNIQUES discussed above, there are other methodologies that can be explored. A way to combine multiple base classifiers of different kind, known as *stacked generalisation*, has been proposed by Wolpert.[12] Some of the steps applied in stacked generalisation are the same as those in cross-validation: For a 2-fold case, we need to split our training data into two disjoint parts. We first train the base classifiers on the first part, and test them on the second one. We use the predictions from the last step as inputs for training a higher level learner.

[12] Wolpert, D. H. (1992). Stacked generalization. *Neural Networks* 5(2), 241–259

See our discussion of cross-validation in Section 3.12.

An alternative name for stacking is *blending* and the word was made popular by the winners of the famous Netflix Prize Competition to improve the accuracy of recommendations provided by Netflix, based on customer film preferences[13]. Sometimes a distinction between the two is introduced: In blending we create a small holdout set out of the training set, and the stacker model is trained on the holdout set. In general, blending is a simpler methodology than stacking, but less data is actually used. Please note that the stacker may overfit to the holdout set.

Blending is another altenative.

[13] Töscher, A. and M. Jahrer (2009). The BigChaos solution to the Netflix grand prize. http://www.netflixprize.com/assets/GrandPrize2009_BPC_BigChaos.pdf

7.4 Ensemble Techniques in Action

LET US NOW TURN OUR attention to the Scikit-learn implementation of the various ensemble techniques we have covered. We will continue working with the Titanic dataset we used in Section 7.1.1.

We will continue using the Titanic dataset.

We will apply the same pre-processing steps to the training set as we did before, but this time we are going to add more variables, including the port of embarkation. This feature will need to be transformed in order to obtain dummy variables to encode the three categories:

We apply the same pre-processing as before.

- Cherbourg: C

- Queenstown: Q

These are the three levels for the port of embarkation.

- Southampton: S

For completeness, let us detail the steps again:

```
titanic = pd.read_csv(u'./Data/train.csv')


titanic = titanic.drop(['Ticket','Cabin'], axis=1)
titanic = titanic.dropna()


titanic = pd.concat([titanic,\
pd.get_dummies(titanic['Sex'])], axis=1)


titanic = pd.concat([titanic,\
pd.get_dummies(titanic['Embarked'])], axis=1)
```

These steps are similar to those applied in Section 7.2.1.

We need to encode the port of embarkation.

Let us split our dataset into training and testing. We will use the former to train various models, and the latter for evaluation:

```
import sklearn.model_selection as ms


X = titanic[['Pclass','Age','female','SibSp',\
'Parch','Fare','S','C','Q']]
Y = titanic['Survived']


XTrain, XTest, YTrain, YTest =\
ms.train_test_split(X, Y,\
test_size= 0.2, random_state=42)
```

We construct a matrix with the required features.

We then create our training and testing datasets.

We are interested in employing a variety of ensemble methods implemented in Scikit-learn, namely:

- Bagging: `BaggingClassifier()`

- Boosting: `AdaBoostClassifier()`

- Random Forests: `RandomForestClassifier()`

- Extremely Randomised Trees: `ExtraTreesClassifier()`

These ensemble techniques in Scikit-learn accept a parameter called `base_estimator` that determines the base classifier to be used in modelling. By default, the base classifier is taken to be a decision tree, such as `DecisionTreeClassifier`, which we have discussed in Section 7.2. They also take a parameter that determines the total number of base classifiers to be used, namely `n_estimators`.

The parameter `base_estimator` determines the base classifier.

The number of base classifiers to use is given by `n_estimators`.

Let us import the relevant modules we will use:

```
from sklearn.metrics import roc_curve

from sklearn.tree import DecisionTreeClassifier

from sklearn.ensemble.weight_boosting import \
AdaBoostClassifier

from sklearn.ensemble import BaggingClassifier

from sklearn.ensemble.forest import \
(RandomForestClassifier, ExtraTreesClassifier)
```

We will use a decision tree as our base classifier.

We will use implementations for Ada-Boost, Bagging, Random Forests and Extremely Randomised Trees.

We can write a script that considers each of the ensemble methods above, as well as a standard decision tree to train a variety of models. Let us prepare the ground for the script by building objects to hold relevant information. For example, we define the number of base classifiers as follows:

```
n_estimators = 100
```

We will use 100 trees in our ensembles.

Let us now define a list with the different models we are going to train:

```
models = [DecisionTreeClassifier(max_depth=3),\
BaggingClassifier(n_estimators=n_estimators),\
RandomForestClassifier(n_estimators=n_estimators),\
ExtraTreesClassifier(n_estimators=n_estimators),\
AdaBoostClassifier(n_estimators=n_estimators)]
```

Then we create a list of models to be trained.

Let us also create a list with the names of the models we are going to use:

```
model_title = ['DecisionTree', 'Bagging',\
'RandomForest', 'ExtraTrees', 'AdaBoost']
```

This list will provide an easy way to identify each of the models.

We also need a suitable structure to keep track of the predictions, probabilities and scores, as well as the true and false positive rates and thresholds:

```
surv_preds, surv_probs, scores,\
fprs, tprs, thres = ([] for i in range(6))
```

We are initialising six empty lists that will get populated next.

We can now traverse the models list , and fit each of our models with the XTrain and YTrain arrays. We can then proceed to obtain predictions and probabilities using XTest and YTest:

```
for i, model in enumerate(models):
    print(''Fitting {0}''.format(model_title[i]))

    clf = model.fit(XTrain,YTrain)
    surv_preds.append(model.predict(XTest))
    surv_probs.append(model.predict_proba(XTest))
    scores.append(model.score(XTest, YTest))

    fpr, tpr, thresholds = roc_curve(YTest,\
    surv_probs[i][:,1])
    fprs.append(fpr)
    tprs.append(tpr)
    thres.append(thresholds)
```

Let us print the name of the model being trained.

We proceed to train the model and calculate the predictions.

Finally we calculate the true and false positive rates and thresholds.

In this case, the scores that we have obtained are as follows:

```
> for i, score in enumerate(scores):
      print(''{0} with score {1:0.2f}''.\
      format(model_title[i], score))


DecisionTree with score 0.75

Bagging with score 0.73

RandomForest with score 0.76

ExtraTrees with score 0.78

AdaBoost with score 0.81
```

We can now print the scores obtained when training each of the models in question.

Notice that we have not performed any *k*-folding in our cross validation, and you are encouraged to do so. For the time being, we can obtain a wider view of the performance of these models by looking at the ROC.

In Figure 7.9 we have the ROC curves, and their corresponding AUC scores, obtained with the various ensemble models used. We can see that all of the models used perform much better than random guessing (diagonal dotted line). For the simple splitting we have made, the ROC curves indicate that the ensemble models also perform better than the decision tree. Ada-Boost (AUC = 0.83) and Random Forest (AUC=0.82) are seen to be better at this classification task than the rest.

Remember that the decision tree is our base classifier.

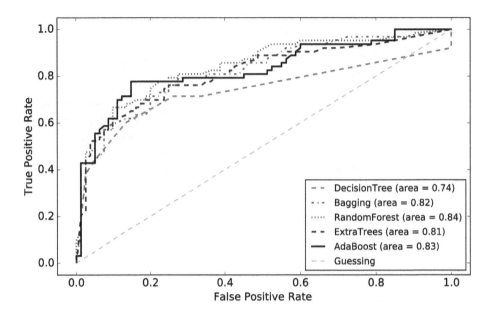

Figure 7.9: ROC curves and their corresponding AUC scores for various ensemble techniques applied to the Titanic training dataset.

7.5   Summary

THE INSPIRATION THAT TREES HAVE provided in the endeavour of organising and visualising information is undeniable. Their appeal is palpable in different disciplines as relevant communication tools, depicting information about biological species, family relationships, database schemas or computational algorithms. In this chapter we have seen how trees have also influenced the data science arena with algorithms in both supervised and unsupervised learning thanks to the hierarchical organisation that a tree provides.

In regards to unsupervised learning, we considered hierarchical clustering. In particular we concentrated in the description of agglomerative clustering, where we build a hierarchy from the bottom-up. Starting with data instances as their own cluster, we successively merge similar clusters as we go up in the levels of the tree. A way to visualise the result of hierarchical clustering is with the aid of a dendrogram. A dendrogram is therefore a tree-like structure that lets us visualise the clusters obtained with the algorithm. The length of the clades or branches of the dendrogram are related to the similarity between clusters.

We also covered the supervised learning task of classification with the help of decision trees. Decision trees are well-known tools in areas such as operations research or decision analysis. Their diagrammatic representation in terms of rules that guide us through the structure is readily understood, even without knowing how the tree itself was constructed. We discussed a heuristic to grow decision trees without having to construct rules for all the different combinations of features in our dataset. The key is the measure of purity before and after data is split by the application of a rule. We considered the application of entropy, Gini impurity and classification error as measures of purity.

Finally, we appealed to the wisdom of base classifiers to improve on our predictions. Ensembles of these base classifiers result in more accurate results than those obtained with a base classifier on its own. These base classifiers, or weak learners, are required to be more accurate than random guessing and exhibit diversity in their misclassifications, as well as being uncorrelated with each other. Ensemble classifiers can be constructed by manipulating a) the training dataset, b) the class labels, or c) the learning algorithm. We finished this chapter with details about ensemble techniques such as bagging, boosting and random forests.

8

Less is More: Dimensionality Reduction

THE DRAMATIC INCREASE IN DATA we have seen in recent times not only encompasses an explosion in the number of records, but also in the number of features that describe our datasets. In Chapter 1 we alluded to the fact that the richness of the data available to businesses and researchers provides us with opportunities to make better decisions based on the data itself. Nonetheless, sometimes the number of available features can be overwhelming, to the point that it is difficult to determine what attributes of a dataset are the most important ones.

Learn from Mr Creosote: Even if it is wafer thin, an extra mint can be far too much.

We have already discussed the idea behind feature selection as a useful tool under our jackalope data scientist belt, where less is truly more: A careful selection of the features to be included in our models has a large effect in the outcomes of many machine learning algorithms, and in some cases may even help in our understanding of the results obtained. We also saw in Section 4.9 how regularisation techniques, such as LASSO, provide a way to

See Section 3.6 for a discussion on feature selection.

Careful selection of features may have a large effect in our results.

perform feature selection in an automatic way, directly as a result of suitable regularisation measures.

Remember that LASSO allows for regression coefficients to be shrunk down to zero.

In this chapter we we will discuss Principal Component Analysis (PCA) and Singular Value Decomposition (SVD) as ways reducing the number of features in our dataset by generating combinations of the given attributes with the purpose of projecting into a lower dimensional space while maintaining as much information as possible.

8.1   Dimensionality Reduction

IN THE ALGORITHMS AND APPLICATIONS we have been discussing we organise our data in terms of matrices (large and small). This representation enables us to use linear algebra to carry out matrix operations and use compact notations to express our models. For instance, in Section 3.10 we saw how Scikit-learn expects data to be represented by numeric matrices with M instances (row-by-row) and N features in the columns.

We have been representing our data in terms of matrices.

We can think of the features in our data as the different dimensions along which our data instances are specified. We are also aware of the so-called *curse of dimensionality*, where the increase in the number of dimensions requires us to have a much larger number of instances in our datasets. A way to deal effectively with this problem is by considering the features that we are including in our models. A careful selection of these features goes a long way, and sometimes appropriate transformations have the advantage

See Section 3.9 for a discussion about the curse of dimensionality.

Dimensionality reduction is in a way a form of feature extraction.

of reducing the number of features while conveying as much useful information as possible. Dimensionality reduction is thus a form of feature extraction.

In that respect, the process of dimensionality reduction involves taking datasets represented in large matrices and finding "narrower" ones that are close to the original matrix. These narrower matrices have a smaller number of columns and of rows, and in a way are easier to manipulate than the large matrix we started with. Having fewer dimensions in our problem may improve generalisation, we gain some speed in the running of algorithms and use less storage to hold the data. In this respect, a good example to bear in mind is data compression. For instance, in image compression we are interested in reducing the size of the data while still being able to tell what is depicted. We need to balance file size with the resolution we want to keep.

> Narrower in the sense that having less features leads to less columns in our matrices.

> Improved generalisation leads to less chance of overfitting.

> Image compression is a good example of dimensionality reduction.

The process of finding the "narrower" matrix we are interested in involves the decomposition our original large matrix into simpler meaningful pieces. In many cases, the process involves the calculation of the eigenvalues and eigenvectors of the original matrix, i.e. eigenvalue decomposition. It is worth exploring the importance of the process of eigenvalue decomposition and its meaning.

> Matrix decomposition is an important step in dimensionality reduction.

The mathematician David Hilbert[1] was the first person to use the word *eigen* in this context. The word comes from the German language and is a prefix that can be translated as "proper", "distinct", "own" or "particular". Given an $n \times n$ square matrix $\mathbf{A}$, and a column vector $\mathbf{x}$ with n non-zero

[1] Hilbert, D. (1904). Grundzüge einer allgeminen Theorie der linaren Integralrechnungen. (Erste Mitteilung). *Nachrichten von der Gesellschaft der Wissenschaften zu Göttingen, Mathematisch-Physikalische Klasse, 49–91*

elements we can carry out the matrix multiplication $\mathbf{Ax}$. We can ask the following question: Is there a number λ such that the multiplication $\lambda\mathbf{x}$ gives us the same result as $\mathbf{Ax}$? In other words:

$$\mathbf{Ax} = \lambda\mathbf{x}. \qquad (8.1)$$

If such a number λ exists, then we say that it is an eigenvalue of the matrix $\mathbf{A}$, and $\mathbf{x}$ is one of its eigenvectors. The interest in these quantities comes into place when we start pondering the application of linear transformations of the matrix $\mathbf{A}$.

In this context λ is the eigenvalue of the matrix $\mathbf{A}$.

A linear transformation is a function between two vector spaces that preserves the operations of addition and scalar multiplication. In simpler terms, a linear transformation takes, for example, straight lines into straight lines or to a single point, and they can be used to elucidate how to stretch or rotate an object.

We can think of linear transformation as ways to stretch or rotate an object.

The use of eigenvalues and eigenvectors comes into place as they make linear transformations easier to understand. Eigenvectors can be seen as the "directions" along which a linear transformation stretches (or compresses), or flips an object, whereas eigenvalues are effectively the factors by which such changes occur. In that way, eigenvalues characterise important properties of linear transformations, for example whether a system of linear equations has a unique solution, and they can also describe the properties of a mathematical model.

Eigenvectors can be seen as the direction along which stretching or flipping is applied.

Notice that Equation (8.1) implies that we can recover the result of the matrix multiplication by multiplying the eigenvalues with the appropriate eigenvectors. Once we have decomposed our matrix, the pieces can then be used in the modelling steps of other algorithms. In the case of dimensionality reduction, the decomposition is usually applied as a pre-processing step to get a better idea of the data we are dealing with, and typical examples are in effect unsupervised learning tasks.

This is a much easier operation as eigenvalues are scalars, i.e. numbers.

Dimensionality reduction is a pre-processing step.

The learning objective of dimensionality reduction is the use of data in the most meaningful basis possible. This can be achieved by choosing a subset of features and create new ones out of them. In an unsupervised manner, we consider only the data points themselves and not their labels (if they exist). We look therefore to remove data that is not informative, enabling us to get better generalisation and model performance on new data.

Think about this as the more meaningful coordinates if you will.

Dimensionality reduction is an unsupervised task.

For a dataset represented by a matrix $\mathbf{B}$ with M records and N features, we are interested in finding a d-dimensional representation of $\mathbf{B}$, with $(d \ll N)$ that encapsulates the information in the original dataset based on particular criteria. Linear dimensionality reduction is based on the concept of performing a linear projection of the data as:

We are interested in finding a smaller representation of a matrix while keeping as much information as possible.

$$\mathbf{b} = \mathbf{U}^T\mathbf{B}, \qquad (8.2)$$

where $\mathbf{b}$ has dimensionality d and is a projection of the original matrix $\mathbf{B}$. The projection is obtained with the use of the $N \times d$ matrix $\mathbf{U}$ that defines a d-dimensional linear subspace. We end up with different dimensionality reduction methods depending on the definition of $\mathbf{U}$.

Dimensionality reduction can be seen as a projection of the original matrix.

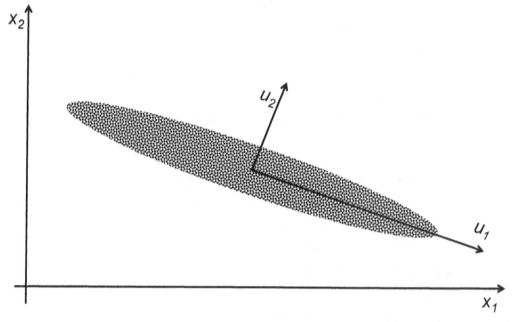

Figure 8.1: A simple illustration of data dimensionality reduction. Extracting features $\{u_1, u_2\}$ from the original set $\{x_1, x_2\}$ enables us to represent our data more efficiently.

Before we enter into further details, let us provide an illustration of the usage of dimensionality reduction. For simplicity, let us consider data in a 2-dimensional space with features $\{x_1, x_2\}$ as shown in Figure 8.1. If we were to carry out feature selection and concentrate only on feature x_1, we naïvely would have indeed reduced the dimensionality of the datase. However, we would need to ask ourselves whether we have lost meaningful information by disregarding x_2.

Here we are naïvely reducing the dimensionality from 2D to 1D.

In view of the representation given in Figure 8.1 we can see that the data has substantial variance along both features x_1 and x_2. Concentrating only on x_1 has the effect of a large loss of information.

Let us now consider a coordinate transformation into the features $\{u_1, u_2\}$ shown in Figure 8.1. These new features are given as a linear combination of the original set $\{x_1, x_2\}$ such that:

$$u_1 = a_1 x_1 + a_2 x_2, \qquad (8.3)$$

$$u_2 = b_1 x_1 + b_2 x_2. \qquad (8.4)$$

The new features $\{u_1, u_2\}$ are given as a linear combination of $\{x_1, x_2\}$.

In this new representation, we can concentrate our attention in the extracted feature u_1. In that case, we are indeed reducing the dimensionality (from 2D to 1D), and most importantly the information loss incurred by ignoring u_2 is much less than in the previous case. This is because the spread of the data is larger along u_1 than along u_2. In other words, there is less variance along the latter feature.

The information loss by concentrating only on u_1 is much less than in the previous case.

We can now use the new representation of our data and concentrate only on feature u_1. We can also apply other algorithms: Linear regression, clustering, decision trees, etc. It is worth emphasising that the new feature u_1 is now a combination of the original features x_1 and x_2 as shown in Equation (8.4). Depending on the nature of the data, the interpretation of the extracted feature u_1 may or may not be straightforward, but there is no doubt that the modelling part will turn out to be much more efficient.

Remember that the extracted feature is a combination of the original features.

8.2 *Principal Component Analysis*

PRINCIPAL COMPONENT ANALYSIS, OR PCA, is a favoured technique for dimensionality reduction. It is widely used among other things for exploratory analysis, data compression and feature extraction. It has application range from brain topology[2] to seasonality analysis[3]. It makes use of an orthogonal transformation to take the original N coordinates of our dataset into a new coordinate system known as the **principal components**. The aim of the technique is to use a reduced subspace, provided by the principal components, that seeks to maintain most of the variability of the data.

[2] Duffy, F. H. et al. Unrestricted principal components analysis of brain electrical activity: Issues of data dimensionality, artifact, and utility. *Brain Topography* 4(4), 291–307
[3] Rogel-Salazar, J. and N. Sapsford (2014). Seasonal effects in natural gas prices and the impact of the economic recession. *Wilmott* 2014(74), 74–81

The principal components are ranked, with the first principal component accounting for the largest contribution to the variance. In succession, each of the rest of the principal components provide their contribution to the variability of the dataset. Since the transformation is orthogonal, each of the components is uncorrelated with each other. The orthogonality constraint comes from the fact that the principal components are the eigenvectors of the covariance matrix. The dimensionality reduction comes into place when representing our data using only those principal components that provide the highest contributions to the variance of the dataset.

The principal components are ranked in decreasing order of contribution to the variance.

We reduce the dimensionality by considering only those principal components with the highest contribution.

As we mentioned above, PCA of a matrix $\mathbf{A}$ is equivalent to performing an eigenvalue decomposition of the covariance matrix of $\mathbf{A}$. A covariance matrix provides a summary of the extent to which corresponding elements from two sets of ordered data move in the same direction. Principal component analysis can be applied to a matrix of any dimensions, but remember that the covariance matrix is a symmetric.

PCA is equivalent to an eigenvalue decomposition of the covariance matrix.

A symmetric matrix is a square matrix that is equal to its transpose.

For a matrix $\mathbf{A}$, it is possible to calculate the mean value of each of its columns and create the vector μ. We define the covariance matrix as:

$$\mathbf{V} = E\left[(\mathbf{A} - \mu)(\mathbf{A} - \mu)^T\right],\qquad(8.5)$$

The covariance matrix can be expressed in this way.

the V_{ij} element of the matrix corresponds to the covariance of the i and j columns of $\mathbf{A}$:

$$V_{ij} = E\left[(A_i - \mu_i)(A_j - \mu_j)^T\right] = \sigma_{ij}.\qquad(8.6)$$

This is the expression for the covariance.

In particular, the diagonal elements of the matrix $\mathbf{V}$ are the variances of the components of $\mathbf{A}$:

$$V_{ii} = E\left[(A_i - \mu_i)^2\right] = \sigma_i^2.\qquad(8.7)$$

This is the expression for the variance.

The eigenvalues of the covariance matrix $\mathbf{V}$ are all real and positive, and the eigenvectors that correspond to distinct eigenvalues are orthogonal. This means that we can decompose the matrix $\mathbf{V}$ as:

$$\mathbf{V} = \mathbf{Q}\mathbf{\Lambda}\mathbf{Q}^T = \sum_{i=1}^{n} \lambda_i \vec{q}_i \vec{q}_i^T,\qquad(8.8)$$

The covariance matrix can be decomposed in this way.

where $\mathbf{Q}$ are the eigenvectors of $\mathbf{A}$, and the values in the matrix $\mathbf{\Lambda}$ are the eigenvalues associated with $\mathbf{A}$.

The covariance matrix provides us with information about the spread (variance) and orientation of our dataset. If we are interested in the representation of our data that provides that largest variability, we simply have to find the direction of the largest spread of data. The largest eigenvector of the

The covariance matrix gives us information about the variance of our dataset.

covariance matrix will point into the desired direction, and its eigenvalue is related to the amount of variance explained by this component. In turn, the second largest eigenvector is orthogonal to the first one, and points in the direction of the second largest variability in the data, and so on.

The first eigenvector points in the direction of the largest variability.

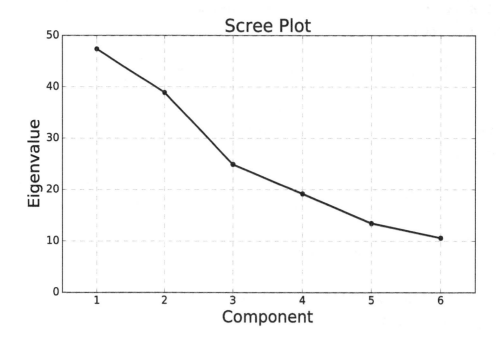

It is usual to represent the eigenvalue associated with a component in a scree plot. We show a diagrammatic example in Figure 8.2 for 6 components and their corresponding eigenvalues. The scree plot may provide an indication of how many principal components to keep. Ideally the scree plot shows a steep downward curve: The number of components to keep are those in the steepest part of the plot. This is sometimes known as the *elbow test*.

Figure 8.2: A diagrammatic scree plot showing the eigenvalues corresponding to each of 6 different principal components.

A scree plot is used to represent the variance contribution of each of the components.

8.2.1   PCA in Action

WE WILL APPLY PCA TO the image of a jackalope[4] shown
in Figure 8.3. It is available at `https://dx.doi.org/10.`
`6084/m9.figshare.2067186.v1`. We will see how principal
component analysis can be used in image compression: We
will take an increasing number of components to reconstruct
the image. We can read the PNG file that into Python with
the help of matplotlib:

[4] Rogel-Salazar, J. (2016b,
Jan).   Jackalope Image.
10.6084/m9.figshare.2067186.v1

```
%pylab inline
from numpy import mean, size
A = imread(r'JRogel_Jackalope.png')
```

We use imread to read an image
from a file into an array.

The size is given by the shape of the matrix A:

```
> print(shape(A))
(1880, 1860, 4)
```

The image has 1860×1880 pixels.

As we can see, the image is made out of 1860×1880 pixels
arranged in four stacked arrays. Each of the first three
arrays corresponds to the red (R), green (G) and blue (B)
channels of the image, and the fourth one is the so-called
alpha, or transparency. In this case, since the image is black
and white, the RGB values would be similar, and for the
sake of simplicity we will work with a single array made
out of the mean value of the 4 arrays provided:

The stacked arrays that make the
image correspond to RGB and
alpha values.

```
A1 = mean(A,2)
```

We take the mean value of the
layers.

We know that the total number of components is given by
the original data set:

Figure 8.3: A jackalope silhouette
to be used for image processing.

```
> full_pc = size(A1, axis=1)
> print(full_pc)
 1860
```

Here we are taking the size of
the matrix along the columns, i.e
axis=1.

We are interested to see if we can reconstruct the image
with a smaller number of components obtained with PCA.
We will write a script that uses a number of components
less than full_pc (in this case 1860) and visually compare
the reconstructed image with the original one.

We want to reconstruct the
image using a smaller number
of components.

The PCA decomposition can be carried out with the implementation in the `decomposition` module in Scikit-learn:

The decomposition module is required.

```
from sklearn import decomposition
```

The `.PCA()` function takes a parameter called `n_components` that tells the algorithm the number of components to keep. If the parameter is not provided, all components are kept. In this case, we are going to see the effects that the image reconstruction has when keeping up to 500 components. First we will fit a PCA model with a determined number of components to keep using the `fit` method. Then we will reconstruct the image using the `inverse_transform` method:

We can tell PCA how many components to keep with the `n_components` parameter.

```
components = range(0,600,100)
fig=plt.figure()
for i, num_pc in enumerate(components):
    i+=1
    pca = decomposition.PCA(n_components=num_pc)
    pca.fit(A1)

    Rec = pca.inverse_transform(pca.transform(A1))
    ax  = fig.add_subplot(2,3,i,frame_on=False)
    # removing ticks
    ax.xaxis.set_major_locator(NullLocator())
    ax.yaxis.set_major_locator(NullLocator())
    imshow(Rec)
    title(str(num_pc) + ' PCs')
    gray()
```

We will programmatically take a number of components to reconstruct the image.

The decomposition is carried out by the `fit` method.

The reconstruction is done with the `inverse_transform` method.

We get rid of the ticks in the axes and show the result.

The results can be seen in Figure 8.4. It is clear that using even 100 components, the image is definitely recognisable, and by the time we are keeping 500 components we can clearly distinguish some of the finer features in the image, such as the white streaks in the ears and the eye of the jackalope.

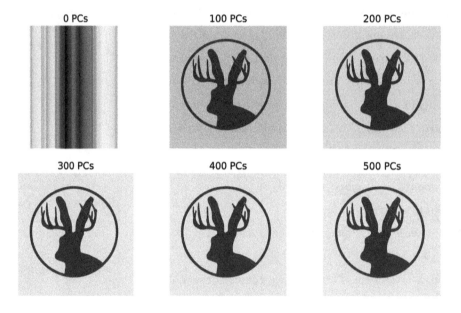

Figure 8.4: Principal component analysis applied to the jackalope image shown in Figure 8.3. We can see how retaining more principal components increases the resolution of the image.

We can take a look at the explained variance ratio we would obtain by allowing all the components in the model and looking at the percentage of variance explained by each of the components:

```
pca1 = decomposition.PCA()
pca1.fit(A1)


var_ratio = pca1.explained_variance_ratio_
```

Remember that by default PCA takes all the components.

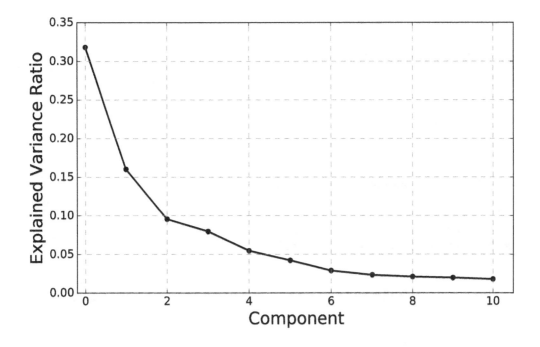

Figure 8.5: Scree plot of the explained variance ratio (for 10 components) obtained by applying principal component analysis to the jackalope image shown in Figure 8.3.

We show a scree plot of these values in Figure 8.5 where we can see that the elbow test would indicate that 8 components are enough to explain the variability in the image.

8.2.2   *PCA in the Iris Dataset*

LET US NOW APPLY PCA to the Iris dataset and use the dimensionality-reduced data in a logistic regression classifier. As usual, let us load the dataset into appropriate arrays:

```
from sklearn import datasets

iris = datasets.load_iris()

X = iris.data
Y = iris.target
```

This should be second nature to us by now.

We start by splitting our data into training and testing datasets:

```
import sklearn.model_selection as ms

XTrain, XTest, YTrain, YTest =\
ms.train_test_split(X, Y,\
test_size= 0.3, random_state=7)
```

We definitely know how to do this and, more importantly, why!

In this case we only have four features: Sepal length, sepal width, petal length and petal width. Let us perform PCA in the training set to give us an idea of the amount of variance explained by the principal components. The result is shown in the scree plot in Figure 8.6. As we can see, perhaps one or two components are good enough to explain the variance in this dataset.

PCA will let us see how many principal components explain most of the variance.

```
from sklearn import decomposition


IrisPCA=decomposition.PCA()

Iris_Decomp = IrisPCA.fit(XTrain)

X_Decomp = Iris_Decomp.transform(XTrain)

var_ratio = IrisPCA.explained_variance_ratio_
```

We simply apply the PCA decomposition to the data. Remember to fit and transform the model.

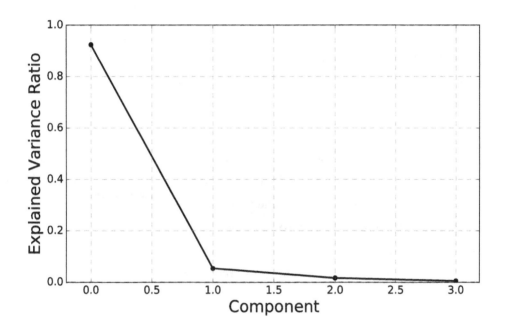

Figure 8.6: Scree plot of the explained variance ratio obtained by applying principal component analysis to the four features in the Iris dataset.

It is possible to use PCA first to extract the features we are interested in and use them in our classification task later; in this case we are employing logistic regression. Let us explore how we can use Scikit-learn to chain the feature extraction to the logistic regression estimator, and at the same time run a search for the best parameters to use. We can do all this with the help of `Pipeline` which sequentially applies a list of transformations and estimators. For our purposes we need to chain `PCA` and `LogisticRegression`:

```
from sklearn import linear_model
from sklearn.pipeline import Pipeline


logistic = linear_model.LogisticRegression()
pca = decomposition.PCA()


pipe = Pipeline(steps=[('pca', pca),\
('logistic', logistic)])
```

The `Pipeline` command enables us to chain models. In this case we chain principal component analysis with a logistic regression.

The steps in the pipeline indicate the transformations that we are expected to carry out. We are also providing a name (included as a string) to be used to refer to the various steps in the pipeline. With that in place, we will conduct a search for both the number of components to keep (from PCA), and the inverse of the regularisation strength used in logistic regression.

The steps in the pipeline indicate the models to be executed.

See Section 6.3.2 for a discussion of logistic regression.

```
from sklearn.model_selection import GridSearchCV


n_components = list(range(1,3))
Cs = np.logspace(-2,4,100)
```

We will search for the best parameters for PCA and logistic regression.

We can now carry out the search employing k-fold cross-validation with 10 folds:

```
Iris_cls = GridSearchCV(pipe,\
dict(pca__n_components=n_components,\
logistic__C=Cs), cv=ms.KFold(n_splits=10))
```

We need to pass the pipeline to the `GridSearchCV` command.

We are ready to use our pipe to fit the model:

```
> Iris_cls.fit(XTrain, YTrain)
> print(Iris_cls.best_params_)

{'pca__n_components': 2,
 'logistic__C': 114.97569953977356}
```

Training the piped model works in the same way as before.

As we can see, the search indicates that 2 components with a value of C equal to 114.98 provides the best estimator. Let us use the result to obtain our prediction:

```
y_pred = Iris_cls.predict(XTest)
```

And the predictions are extracted in the same manner.

Finally, we can obtain a confusion matrix for the prediction of our classifier against the testing set:

```
> from sklearn.metrics import confusion_matrix
> confusion_matrix(YTest,y_pred)

array([[12,  0,  0],
       [ 0, 13,  3],
       [ 0,  3, 14]])
```

In this case we have a three-way confusion matrix.

8.3   *Singular Value Decomposition*

FINDING THE PRINCIPAL COMPONENTS THAT characterise
our data is a powerful tool to have in our jackalope data
science toolbox. However, PCA may not be suitable when
the initial dimensionality of our data is very large. In the
example of the image processing we provided in Section
8.2.1, the resolution of the image could have been very
large: For images in megapixels we have $N \geq 10^6$ and the
covariance matrix will be $\geq 10^{12}$. Not an impossible task per
se, but perhaps other methods may be more suitable and
efficient.

In cases where the dimensionality
of our data is very large, PCA may
not be suitable.

One such method is the *Singular Value Decomposition* or
SVD for short. It has the advantage of offering an exact
representation of any matrix and, most importantly for us,
enables the elimination of parts of the data that are deemed
to be less important. It therefore creates an approximate
representation with the number of dimensions we choose.
In other words, it is a suitable alternative for dimensionality
reduction, and for other applications as we shall discuss
later in this chapter.

Singular Value Decomposition
(SVD) is a good alternative.

Let us consider an $M \times N$ matrix **A**. We define the rank r of
the matrix **A** as the maximum number of linearly
independent rows (or columns). We can decompose the
matrix **A** as:

$$\mathbf{A} = \mathbf{U\Sigma V}^T, \tag{8.9}$$

where **U** is an $M \times r$ column-orthonormal matrix, **Σ** is an
$r \times r$ diagonal matrix whose elements are called the **singular**

A set of vectors are linearly
independent if no vector in the set
is a linear combination of the other
vectors.

Column-orthonormal: each
column is a unit vector & the dot
product of any two columns is
zero.

values of **A**, and finally **V** is an $N \times r$ column-orthonormal matrix. The columns of **U** and **V** are called the **singular vectors** of the matrix **A**. In Figure 8.7 we show a diagrammatic representation of the singular value decomposition.

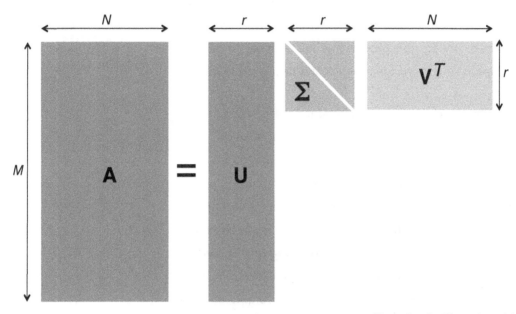

Figure 8.7: An illustration of the singular value decomposition.

The singular vectors of **A** provide orthonormal bases for $\mathbf{AA}^T$ and $\mathbf{A}^T\mathbf{A}$. The SVD process consists of calculating the eigenvalues and eigenvectors of $\mathbf{AA}^T$ and $\mathbf{A}^T\mathbf{A}$. The singular values in Σ are then equal to the square roots of the eigenvalues of **U** (or **V**) in descending order. As we can see from the diagram in Figure 8.7, the number of singular values is equal to the rank r of **A**.

The singular values are the square roots of the eigenvalues of **U** (or **V**).

SVD results in a representation of the original data in which the covariance matrix is diagonal, and much easier to

In SVD, the covariance matrix has a diagonal representation.

handle than a full matrix, as it is the case in PCA. So, where is the dimensionality reduction? We can think of this problem as finding an approximation to **A** with a matrix **Ã** whose rank is r. This problem can be solved by setting the smallest of the singular values of **A** to zero. This translates into the elimination of the corresponding rows from both **U** and **V**. It can be shown[5] that besides achieving a reduction in dimensionality, this process minimises the root-mean-square error between **A** and its approximation.

[5] Golub, G. and C. Van Loan (2013). *Matrix Computations*. Johns Hopkins Studies in the Mathematical Sciences. Johns Hopkins University Press

The dimensionality reduction obtained with SVD underlies some techniques used in document analysis such as latent semantic analysis (LSA), where a term-document matrix is used as the basis to obtain linearly independent components. These components can be thought of as hidden or latent concepts in the original data, and hence the name. After the application of SVD, the words are represented by the rows of **U**, whereas the documents by the rows of **V**. Document similarity is extracted by comparing the rows of **VΣ**. A similar application to this can be found in the design of recommendation systems, as we will discuss in Section 8.4.

In Section 6.4.1 we obtained a term-document matrix to be used with the Naïve Bayes Classifier.

8.3.1   SVD in Action

As a demonstration of the power of singular value decomposition, we shall see an application in terms of data compression and noise reduction. Consider the image of a (pixelated) letter *J* as shown in Figure 8.8. As we can see, it is possible to construct the image with 4 different types of

We will see an example using SVD for image compression and noise reduction.

pixel columns. It is conceivable that we can represent the same data in a more efficient way.

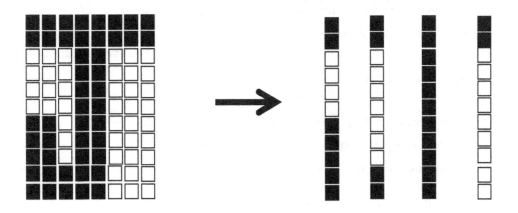

Figure 8.8: An image of a letter *J* (on the left) and its column components (on the right).

Let us construct a noisy picture of the letter *J* shown in Figure 8.8 using 250×150 pixels, i.e. a total of $37,500$ pixels to represent this letter:

```
%pylab inline
import matplotlib.pyplot as plt
import numpy as np


M = np.zeros((250,150))
M[:31,:]=1
M[:,60:91]=1
M[-31:,:60]=1
M[150:,:31]=1
```

We are constructing the pixelated letter *J* with the help of arrays in Python.

We can even add some random noise to the image. The noise can be generated with the help of the `random.uniform` command:

```
M_noisy = np.asmatrix(np.random.uniform(low=0,\
high=0.7, size=(250,150)))


M_noisy = M + M_noisy
```

We are creating a matrix with random entries between 0 and 0.7.

The linear algebra module in NumPy has an implementation of SVD. With it we can readily apply the method to our noisy matrix:

```
U, s, V = np.linalg.svd(M_noisy)
```

NumPy has an implementation of SVD in the linear algebra module.

We can see the first 10 singular values obtained from the operation above in Figure 8.9. It is clear that after the fourth component, the curve flattens out.

Let us now put together a script that takes an increasing number of singular components from 1 to 4, and see if we can reconstruct our image using fewer pixels and even reduce the noise:

```
for S in range(1,5):


    Sig = mat(np.eye(S)*s[:S])


    U_reduced = U[:,:S]
    V_reduced = V[:S,:]


    M_rec = U_reduced*Sig*V_reduced
```

The `mat` command interprets the result as a matrix and eye corresponds to array representing an identity matrix.

We get rid of the columns we do not need.

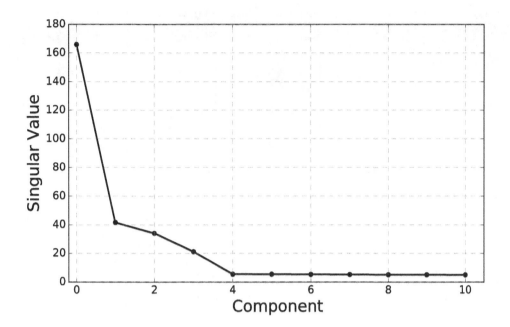

Figure 8.9: The singular values obtained from applying SVD in a an image of a letter J constructed in Python.

The result of the reconstruction process presented above can be seen in Figure 8.10. Each of the panels represents the reconstruction using different number of components. As we can see, the use of 4 singular values renders a very good result.

In terms of the number of elements used in the reconstruction of the image, we needed only $1,604$: With 4 singular values, we have $1,000$ from the 250×4 U_reduced matrix and 600 from the 4×150 V_reduced matrix. That is a good reduction from the original $37,500$ elements in the original, with the added benefit of having lessened the noise in the image, as is clearly seen when comparing the first and last panels in Figure 8.10.

We have reduced the number of pixels needed, as well as the noise.

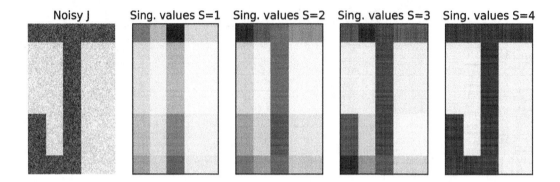

The example above is a simplified application of SVD to more complex images and indeed other data in general. We can consider for instance its use in regression where a clear linear relationship exists between a feature and the target variables, where small noisy errors come into play. Singular value decomposition may enable us to determine the direction in which the data aligns better and safely ignore the rest of the singular values. SVD can be used to detect groupings in the data. This can be useful in determining similarities among data instances. This pattern detection is exploited for example in building recommendation systems and we shall explore some of the most important concepts in the following section.

Figure 8.10: Reconstruction of the original noisy letter J (left most panel), using 1-4 singular values obtained from SVD.

We can apply SVD as a pre-processing step.

Think of the four types of columns that make up our letter J.

8.4   Recommendation Systems

WE ARE ALL FAMILIAR WITH online outlets offering us a variety of products from household items, books, music, films, recipes, friends and partners. Some are more specialised than others and they all have one thing in common: They are interested in continuing a relationship with us, the customers. One way to continue our engagement with these services is by obtaining meaningful recommendations.

A variety of outlets offer us a wide range of services, and recommendations are a good way to engage with us.

An example of the importance of this type of application is the now famous 2006 Netflix competition[6] offering a $1 million USD prize to anyone who could improve their recommendation engine. The 2009 winning entry improved the system by 10%. However, the solution was never implemented due to the engineering efforts that were required.

[6] Töscher, A. and M. Jahrer (2009). The BigChaos solution to the Netflix grand prize. http://www.netflixprize.com/assets/GrandPrize2009_BPC_BigChaos.pdf

Picture then the following situation: You are interested in a streaming service that lets you play videos in your favourite device. You create your account and are presented with some of the most popular offerings the service provides as a form of recommendation. The service provider is interested in predicting the rating you are likely to give to an item you have not rated yet. If the rating is high, then the item can be a good product to put in the recommended list and keep you happy and watching.

The aim of a recommendation engine is to predict the likely rating we would give to an unseen item.

Sounds good, but how is that rating produced? The service provider will use information from the items on offer, as well as from other users and their characteristics to try to predict the rating. In some cases maybe even exploiting social network analysis can be done. There are two main camps for this:

Information about the items and from others users is exploited.

1. **Content-based filtering**, where items are mapped into a feature space using their attributes. The recommendations depend on the characteristics of the items.

   Content-based filtering uses the attributes of the items.

2. **Collaborative filtering** considers data from the ratings that other users have provided for specific items. The recommendations depend on the preferences expressed by the user.

   Collaborative filtering also uses information from other users.

8.4.1   Content-Based Filtering in Action

CONTENT-BASED FILTERING REQUIRES US TO specify the attributes or features that describe the items in our database. We also need to obtain the scores that our users give to each of these features. We can then represent users and items in terms of vectors in the feature space. The item vectors provide a measure of the degree to which items are described by each of the features in question. Similarly, user vectors measure the preferences that the users have for each of the features. We can assume that users will prefer items that are *similar* to the preferences they have expressed.

Content-based filtering is based on the use of vectors describing items based on a series of attributes.

User vectors tell us the preference that users have for the given attributes.

From the discussion about similarity measures in Section 3.8, we know that a convenient way to define similarity is provided by the *cosine similarity*. In this case, the ratings can be obtained by taking the dot product of the user and item vectors and divide by the product of their norms. Let us go through an example that, although is a little bit oversimplified, will let us see the way in which content-based recommendations work. Imagine that we have set up

For a discussion on cosine similarity see Section 3.8.

a streaming service and our catalogue contains a total of 5 films: *Cronos* (1993), *Life of Brian* (1979), *The Never-Ending Story* (1984), *Pinocchio* (1940), and *Titanic* (1997).

Our streaming service has a rather limited catalogue, but will help us see how content-based filtering works.

The feature space we will consider is given by revenue at the box office, film suitability for children, and Oscar winning film. The scores (1-5) for these features are shown in Table 8.1.

ID	Film	Box Office	For Children	Oscar
1	*Cronos*	2	1	1
2	*Life of Brian*	3	3	1
3	*The Never-ending Story*	2	5	1
4	*Pinocchio*	3	5	5
5	*Titanic*	5	2	5

Table 8.1: Films considered in building a content-based filtering recommendation system.

ID	Film	Box Office	For children	Oscar
a	Graham	4	2	5
b	Terry G.	1	2	2
c	Eric	5	4	4
d	John	4	3	1
e	Michael	3	2	5
f	Terry J.	1	1	4

Table 8.2: Scores provided by users regarding the three features used to describe the films in our database.

We can now look at the preferences that our (Pythonic) users have expressed for the three features describing the films in our database. The user scores are shown in Table 8.2. Let us calculate the rating that *Graham* is predicted to give to *Cronos* by obtaining the cosine similarity of the user and film score vectors:

$$
\begin{aligned}
Sim_{Graham-Cronos} &= \frac{(4)(2) + (2)(1) + (5)(1)}{\sqrt{4^2 + 2^2 + 5^2}\sqrt{2^2 + 1^2 + 1^2}}, \\
&= \frac{15}{\sqrt{(45)(6)}}, \\
&= 0.9128. \qquad\qquad (8.10)
\end{aligned}
$$

We calculate the cosine similarity between the user and item vectors.

It seems that *Cronos* is indeed a good match for *Graham*. We show below a function that receives as input information from films and users stored in pandas dataframes.

```
import numpy as np, sys
def content_recomm(user, user_df, film_df):
    try:
        u = user_df.loc[user][1:].values
    except:
        print(''Error: User does not exist '',\
        sys.exc_info()[0])
        sys.exit(1)
    u_norm = np.linalg.norm(u)
    film_recom = []
    for row in range(shape(film_df)[0]):
        f_name = film_df.index[row]
        f = film_df.ix[:,1:].iloc[row].values
        f_norm = np.linalg.norm(f)
        s = np.dot(u, f)/(u_norm*f_norm)
        if s>0.8:
            film_recom.append((f_name, s))
    film_recom = sorted(film_recom,\
    key=lambda x: x[1], reverse=True)
    return film_recom
```

We check that the user exists, otherwise we terminate execution.

We take the norm of the user vector.

In turn we take each of the item vectors and use them to calculate the cosine similarity.

After checking for the threshold, we sort the recommendations and report them.

In the code above we pass on the name of a user (an ID would be preferable in general, though) and calculate the cosine similarity of the user against the films in the database. We then report only those films for which the cosine similarity is greater than 0.8.

We can decide to use a more stringent threshold.

We apply the function to our dataset as follows:

```python
import pandas as pd

films = pd.read_csv(u'./Data/FilmCB.csv',\
index_col=1)
users = pd.read_csv(u'./Data/UsersCB.csv',\
index_col=1)


r1 = content_recomm('Graham', users, films)
r4 = content_recomm('John', users, films)
r6 = content_recomm('Terry J.', users, films)
```

We are assuming that the data is stored in appropriate csv files.

The content_recomm function can be applied to each individual user.

Let us see the results for *Graham*:

```python
print(r1)

[('Titanic', 0.99401501176863483),
 ('Cronos', 0.9128709291752769),
 ('Pinocchio', 0.91214859859201181)]
```

We have three recommendations for *Graham*.

As we can see, the top three films for Graham in descending order of cosine similarity are *Titanic*, *Cronos* and *Pinocchio*. We can do the same for *John* and *Terry J*. We would recommend to *John* all of the five films starting with *Life of*

Recommendations for other users are obtained in a similar fashion.

Brian (0.9898) followed by *Cronos* (0.9607), whereas for *Terry J.* we only have two recommendations: *Titanic* (0.866) and *Pinocchio* (0.8592).

Content-based filtering works well in cases where there are clear attributes that describe the items to be recommended and for which users have provided clear preferences. However, it has some important drawbacks:

- We need to map each item into a feature space. In general, the feature space may be rather large and the mapping process is very resource-intensive.

  Mapping items into the feature space is resource-intensive.

- The recommendations obtained are very limited in scope as items must be similar to each other.

- User preferences have to be obtained in order for a recommendation to be processed. This is known as the **cold start problem.** It is therefore difficult to provide recommendations to new users who typically have not provided rating information.

  The cold start problem is a significant drawback.

- The nature of the filtering makes it difficult to obtain cross-content recommendations. This is because it requires a comparison of items from potentially different feature spaces.

  Cross-content recommendation is difficult.

8.4.2   *Collaborative Filtering in Action*

THE MAIN ASSUMPTION OF COLLABORATIVE filtering is that users get value from recommendations based on other users with similar preferences and tastes. While in content-based filtering we have used the similarity of items

Collaborative filtering uses a utility matrix to recommend items based on other users' preferences and tastes.

to determine the recommendations, in collaborative filtering instead we make use of a utility matrix whose elements are the preferences reported by users about the items on offer.

There are several approaches that we can take to build our recommendation. For instance, we could take a look at the ratings in our utility matrix to create a matrix detailing the similarity among items. This approach is known as **item-based collaborative filtering** or **memory-based collaborative filtering** and the recommendations that a user receives are based on other items that the user has rated highly in the past. This approach can be thought of in terms of clustering items based on ratings provided by the user. Notice that this item-based filtering is different from content-based filtering as we are not mapping the attributes of the item to a feature space.

There are several approaches to content-based filtering.

In item-based collaborative filtering the user receives recommendations based on items she has rated in the past.

Another alternative is provided by the so-called **model-based collaborative filtering** where instead of looking for similarity among items, we consider the utility matrix to be the result of the product involving two thinner matrices **U** and **V**, encapsulating *latent* concepts in our data. The utility matrix is understood as the ratings that individual users give to specific items available.

Model-based collaborative filtering is based the rating that individual users give to specific items available.

Let us consider the utility matrix shown in Table 8.3 detailing the entries (1-10) that users have provided for a set of books: *I. Robot* (I. Asimov, 1950), *The Martian* (A. Weir, 2011), *Do Androids Dream of Electric Sheep?* (P. K. Dick, 1968), *2001 Space Odyssey* (A. C. Clarke, 1968) and *Solaris* (S. Lem, 1961). The question marks in the table indicate items for

Here we will exemplify collaborative filtering for a set of books.

User	I, Robot	The Martian	Do Androids Dream of Electric Sheep?	2001 Space Odyssey	Solaris
Alice	8	2	10	5	1
Bob	4	?	2	10	9
Carl	3	8	4	9	10
Daniel	5	10	4	9	10
Eve	7	2	9	6	?

Table 8.3: Utility matrix of users v books used for collaborative filtering. We need to estimate the scores marked with question marks.

which users have not provided a rating. The missing values occur because there may be many more users than items, plus users only provide scores for a small portion of the items on offer.

Missing values in the utility matrix are denoted with a question mark.

In this case we are not so much interested in reconstructing the utility matrix as it was the case for the example discussed in Section 8.3.1. Instead we are interested in estimating the missing values in the matrix. Typical applications make use of very sparse matrices.

Our task is to estimate the missing values in the utility matrix.

The idea is that if a user has not rated an item it is likely to be because they have not had a chance to "experience" it. If the estimate is high then the item is a good candidate to be recommended to that particular user. The use of SVD makes it possible to find those missing values without necessarily having to determine all the missing scores as we shall see.

SVD makes this possible.

Remember that SVD decomposes a matrix $\mathbf{A}$ as:

$$\mathbf{A} = \mathbf{U \Sigma V}^T. \tag{8.11}$$

We can think of $\mathbf{U}$ as a matrix where users are represented as row vectors containing linearly independent components. Similarly, the matrix $\mathbf{V}$ corresponds to the items represented as linearly independent row vectors. For the i-th user we can define the row vector problem

$$\mathbf{p}_i = U_i \sqrt{\mathbf{\Sigma}}^T. \tag{8.12}$$

Similarly, for the j-th item we have

$$\mathbf{q}_j = \sqrt{\mathbf{\Sigma}} V_j^T; \tag{8.13}$$

and the score that user i will give to item j is given by:

$$r_{ij} = \mathbf{p}_i \mathbf{q}_j^T. \tag{8.14}$$

This works for full matrices, but in this case we are dealing with sparse ones. If we impute an initial value of zero we are effectively indicating that the user will definitely not prefer the item. Some potential solutions include normalising the ratings by subtracting the mean rating by user for example or, as we will do in this case, taking the mean rating of the item in question.

The objective function that we are trying to minimise can then be expressed as follows:

$$\min_{\mathbf{p}^* \mathbf{q}^*} \sum_{(i,j)} \left(r_{ij} - \mu - \mathbf{p}_i \mathbf{q}_j^T \right)^2 + \lambda \left(|\mathbf{p}_i|^2 + |\mathbf{q}_j|^2 \right), \tag{8.15}$$

where μ is the mean value we referred to above, and λ is the hyperparameter that controls the amount of regularisation.

The matrices $\mathbf{U}$ and $\mathbf{V}$ are related to the users and items.

This matrix decomposition enables us to formulate the problem in terms of an objective function to be optimised, as we shall see below.

We have sparse matrices due to the cold start problem.

In this case we wil use the mean rating for an item to avoid this issue.

The objective function we are trying to minimise is given by the matrix decomposition defined above.

Please note that this optimisation problem has two unknowns, i.e. **p** and **q** and convexity is not guaranteed. A further discussion goes beyond the scope of this book, and you are recommended to read more about methods such as *alternating least squares*[7,8] to tackle this problem.

In this case, we will be tackling this problem with a naïve approach (i.e., use with care!) to find the likely ratings that Bob will give to *The Martian*, and Eve to *Solaris* (see Table 8.3). Starting with a csv file holding the data from Table 8.3 we can load the information into a Pandas dataframe. We still have to deal with the cold start problem as the SVD method will not be able to deal with missing data. In this case we will initialise the missing values with the mean score provided by other users:

[7] Takács, G. and D. Tikk (2012). Alternating least squares for personalized ranking. In *Proceedings of the Sixth ACM Conference on Recommender Systems*, RecSys '12, New York, NY, USA, pp. 83–90. ACM

[8] Hu, Y., Y. Koren, and C. Volinsky (2008). Collaborative filtering for implicit feedback datasets. In *Proceedings of the 2008 Eighth IEEE International Conference on Data Mining*, ICDM '08, Washington, DC, USA, pp. 263–272. IEEE Computer Society

```
import pandas as pd
import numpy as np


A = pd.read_csv(u'./Data/CF_Table.csv',\
index_col=0, na_values=['?'])


A.fillna(A.mean(), inplace=True)
```

We can tell pandas what characters represent missing values with na_values.

We replace missing values with fillna.

In the code above we are assuming that the missing data in the comma-separated-value file is denoted with a question mark, as shown in Table 8.3. We load the data into a Pandas dataframe and specify that any question marks encountered are replaced by NaN so that the software handles the missing data appropriately. In this case we are replacing the missing values by the mean score given to each item with the help

The various methods in Pandas let us manipulate the missing values as required.

of the `.fillna` method. Please note that the replacement is requested to be done in place. In this case, Bob's initial rating for *The Martian* is calculated to be 5.5, and Eve's initial rating for *Solaris* is estimated to be 7.25.

Let us now write a function that uses SVD to improve on the predicted scores:

```
import sys

def cf_recomm(user, book, dat, S=2):
    try:
        uind = dat.index.get_loc(user)
        bind = dat.columns.get_loc(book)
    except:
        print(''Error: User/item doesn't exist'',\
        sys.exc_info()[0])
        sys.exit(1)
    else:
        uind = dat.index.get_loc(user)
        bind = dat.columns.get_loc(book)


    U, s, V = np.linalg.svd(dat)
    Reduced_Sig = np.mat(np.eye(S)*s[:S])
    U_reduced = U[:, :S]
    V_reduced = V[:S, :]
    recom = U_reduced[uind, :]*Reduced_Sig*\
    np.mat(V_reduced[:, bind]).T


    return recom.item(0)
```

We check that the user (or item) exists, otherwise we terminate execution.

The scores can be obtained using only the relevant parts of the utility matrix. There is no need to compute the entire multiplication.

SVD is applied to the data and the dimensionality is reduced. We are using a minimum of $S = 2$ components.

We use the reduced matrices to obtain our estimate and report it.

We can now use the function to determine the scores as follows:

```
> bob_score = cf_recomm('Bob', 'The Martian', A)
> eve_score = cf_recomm('Eve', 'Solaris', A)
> print(bob_score)


7.233127706402881


> print(eve_score)


5.323332988443551
```

We only need to provide the function with the relevant row and column from matrix **A** to calculate the rating estimate.

This results in Bob's score for *The Martian* being 7.23, while Eve is predicted to score *Solaris* with 5.32. We can then recommend Bob to consider reading *The Martian*, whereas Eve may want to consider other items instead of *Solaris*. SVD has enabled us to decompose the utility matrix shown in Table 8.3 into latent features that users rate highly or poorly. In this case for instance, these latent features could be themes such as books with *sentient robots* or *action taking place in Space*.

The singular value decomposition can be understood in terms of latent features in our data.

Please note that in the example above, other than using SVD to find the missing values in our utility matrix, we have not employed any further optimisation or even considered the use of cross-validation. There are a number of recommendation system implementations for Python and we recommend taking a look at some such as `pysuggest`, `crab`, and `python-recsys`.

Pysuggest: `https://code.google.com/p/pysuggest`
Crab: `https://github.com/muricoca/crab`
Python-recsys: `https://github.com/ocelma/python-recsys`

Recommendation systems that use model-based collaborative filtering are very flexible, and there are a few things that need to be taken into account when using them, including:

- Large datasets are needed to construct meaningful utility matrices

- The utility matrices obtained are very sparse

- The possibility of *shilling attacks* must be considered. A shilling attack consists of malicious users providing a large number of very positive ratings to their own products and very negative ones to their competitors', skewing the recommendations obtained

- Initial values for the preferences of a new user are potentially non-existent and therefore a lot of data is needed to tackle this cold start problem. The use of implicit feedback is a way to deal with this issue

Utility matrices can be large and sparse.

Shilling attacks consist of users providing a large number of positive reviews to their own products for example.

Implicit feedback includes information from other sources such as search patterns, browsing history, etc.

8.5   Summary

THE CURSE OF DIMENSIONALITY INTRODUCED in Section 3.9 is an important issue that has to be considered in any data science modelling. This is particularly true in those cases where we have a large number of dimensions, imposing the need for larger datasets. We started this chapter by introducing a couple of techniques for dimensionality reduction that no jackalope data scientist should forgo.

It is clear that using the most important features that describe our data science problem is the best way to proceed in the modelling part of our workflow. Therefore, careful feature selection can be of great help in cases where we have a large number of features. An alternative for this is extracting features. This may help reduce the number of dimensions, while conveying as much useful information as possible.

In Principal Component Analysis (PCA), given a feature matrix $\mathbf{A}$, we are interested in performing an eigenvalue decomposition of the covariance matrix. This is because we want to obtain a representation (orientation) of our data that provides the largest variability. The eigenvalues obtained from PCA are related to the amount of variance explained by each of the associated eigenvectors, or components. This enables us to represent our dataset with a reduced number of features and even tackle issues with noisy data.

Another approach we saw in this chapter is Singular Value Decomposition (SVD) of the matrix $\mathbf{A}$ directly. The method offers an exact representation of any matrix and enables us to eliminate parts of the data that are deemed to be less important. This is done by finding an approximation $\tilde{\mathbf{A}}$ to the original matrix by setting the smallest singular values of $\mathbf{A}$ to zero. This process eliminates the corresponding rows of both $\mathbf{U}$ and $\mathbf{V}$, the component matrices. We also discussed how SVD can be used in building recommendation systems.

In general, dimensionality reduction helps with:

• Reducing computational expense

• Reducing noise in the dataset

- Increasing the generalisation of our predictions

- Reducing overfitting

- Enhancing our understanding of the data and the outcomes of our models.

9

Kernel Tricks up the Sleeve: Support Vector Machines

THE DREADED CURSE OF DIMENSIONALITY can be a tough adversary for any jackalope data scientist out there. Fortunately, we now know that there are a number of tools that can be used to defend ourselves against it: From careful feature selection through to dimensionality reduction. In the last chapter we saw how Principal Component Analysis (PCA) and Singular Value Decomposition (SVD) can be used to reduce the number of features in our problem. This reduction is achieved thanks to combinations of the original attributes that conserve as much information as possible.

Probably as effective as defending yourself agains a man armed with a banana...

Remember that the number of features corresponds to the number of dimensions in our problem.

Another way to tackle our machine learning problems is with the application of suitable transformations to our data. Take for instance the logarithmic transformation we applied in Section 4.5, where we successfully managed to represent our data in terms of a linear relationship in the transformed space. We can now take a step forward and combine the

We would like to combine the task of selecting features with the application of useful transformations.

idea of selecting relevant features by applying suitable transformations.

In this chapter we will consider the use of kernel functions as a way to manipulate datasets in their own original feature space, but as though they were projected into a higher dimensional space. This transformation lets us carry out manipulations in a more straightforward manner.

We then explain the Support Vector Machine (SVM) algorithm.

9.1   Support Vector Machines and Kernel Methods

WE HAVE CONCENTRATED SO FAR in extracting features from our existing data to obtain a representation using fewer dimensions while retaining as much information as possible. A different approach is to transform the data in such a way that our learning tasks look easier to be carried out than in the original feature space.

Instead of extracting new features, we can perform transformations to simplify the learning tasks.

In a sense, this is what we have done when applying a logarithmic transformation to the mammals dataset in Section 4.5: In the transformed feature space, the relationship between a mammal's body and brain is better approximated by a straight line than when using the untransformed features. Once we have carried out the learning task, we need to invert the transformation to the original feature space.

We have done this with a logarithmic transformation in Section 4.5.

Continuing with that train of thought, let us consider the datasets shown in Figure 9.1 using a space spanned by features X_1 and X_2. The data points shown in panel a) can be separated into two groups by using a diagonal line from the upper left-hand corner to the lower right-hand one. In other words, we say that the dataset is linearly separable. However, the squares in panel b) cannot be separated in such an easy manner from the circles in this representation. It may be possible that the data points are actually linearly separable in a different feature space, perhaps even in a higher dimensional one. If that is the case, we can try to find the transformation as we did for the mammals dataset and carry out our classification task in the new space.

Linear separability refers to the possibility of discriminating data classes with a single decision surface.

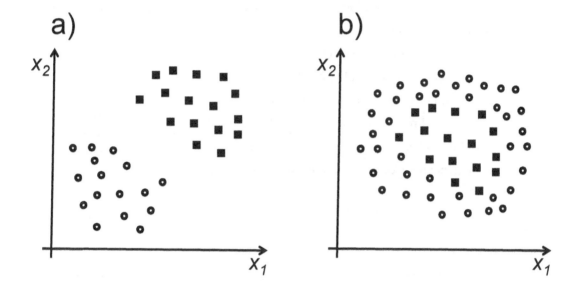

Figure 9.1: The dataset shown in panel a) is linearly separable in the $X_1 - X_2$ feature space, whereas the one in panel b) is not.

Although it is possible to find a new feature space where linear separability can be achieved, there are a couple of things we need to consider. First of all, we may not know the transformation that needs to be carried out, and we will need to search for it. Second, if indeed the data is separable in a higher dimensional space, we are adding complexity to

We may not know the transformation that needs to be applied, plus we are adding complexity to the model.

our model. Could it possible to manipulate the dataset in its original feature space, but as if it were projected into a higher-dimensional one without the need to explicitly carry out the transformation? It sounds far-fetched but it turns out that the answer to that question is *YES*, and we will discuss this point in further detail in Section 9.1.2. However, before we get there, let us consider the linearly separable case.

It is possible though to manipulate the dataset without the need to explicitly carry out the transformation.

As Jorge Luis Borges would remind us[1] a line is made out of an infinite number of ponts, a 2-dimensional space is made out of an infinite number of lines, a 3-dimensional one is made out of an infinite number of planes, etc. As such, a linear classifier in 2D will correspond to a decision boundary given by a line (as discussed above). In 3D it will be a plane, and in higher dimensions we can talk about a hyperplane. In general we can express a linear classifier as:

[1] Borges, J. L. (1984). *El Libro de Arena*. El Ave Fénix. Plaza & Janés

In general we talk about a hyperplane as the decision boundary for a linearly separable dataset.

$$f(\mathbf{x}) = \mathbf{w}^T \mathbf{x} + b, \qquad (9.1)$$

where the weight vector $\mathbf{w}$ is normal to the hyperplane and b is the bias. In our classification task, the training data is used to learn $\mathbf{w}$ and not used again since we can directly use the weight vector for classifying new unseen data.

The weight $\mathbf{w}$ is normal to the hyperplane and b is the bias.

If the data points are linearly separable we can end up with a situation like the one shown in Figure 9.2, where we have more than one separation boundary. Not only is our task to find the classification boundary, but also to find the best one.

Our task is to find the best classification boundary.

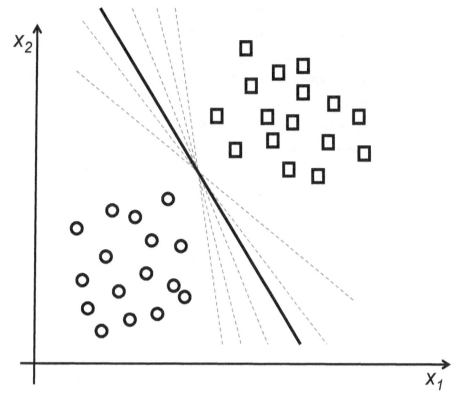

Figure 9.2: A linearly separable dataset may have a large number of separation boundaries. Which one is the best?

9.1.1   *Support Vector Machines*

A SUPPORT VECTOR MACHINE IS a binary linear classifier where the classification boundary is built in such a manner as to minimise the generalisation error in our task. Unlike other classifiers we have discussed, the support vector machine boundary is obtained using geometrical reasoning instead of algebraic. With that in mind, the generalisation error is associated with the geometrical notion of a **margin**, which can be defined as the region along the classification boundary that is free of data points.

A support vector machine minimises the generalisation error associated with the geometrical notion of a margin.

In that manner, a support vector machine (SVM) has the goal of discriminating among classes using a linear decision boundary that has the largest margin, giving rise to the so-called **maximum margin hyperplane** or (MMH). Having the maximum margin is equivalent to minimising the generalisation error. This is because using the MMH as the classification boundary minimises the probability that a small perturbation in the position of a data point results in a classification error. Intuitively, it is easy to see that a wider margin results in having better defined and separate classes.

The maximum margin hyperplane is the linear decision boundary with the largest margin.

A wider margin results in having better defined and separate classes.

Given the discriminant function in Equation (9.1), we can determine the class label of a new record by considering the sign of the function $f(\mathbf{x})$:

$$f(\mathbf{x}_i) = \mathbf{w}^T \mathbf{x}_i + b \begin{cases} \geq 0 \rightarrow y_i = +1, \\ < 0 \rightarrow y_i = -1. \end{cases} \qquad (9.2)$$

The class label of a new record is given by the sign of the discriminant function.

The weight vector $\mathbf{w}$ defines the orientation of the classification boundary, whereas the bias b specifies a translation from the origin. In Figure 9.3 we depict the main components of a support vector machine: The maximum margin is defined by the **support vectors**, i.e. the data points that lie closest to the classification hyperplane. These points are marked in black in Figure 9.3. The task of finding the MMH boils down to the optimisation of a convex objective function and this means that we are guaranteed to obtain a global optimum.

The support vectors are the data points closest to the classification hyperplane.

The support vectors are in fact the data points that are the most difficult to classify as they are closest to the

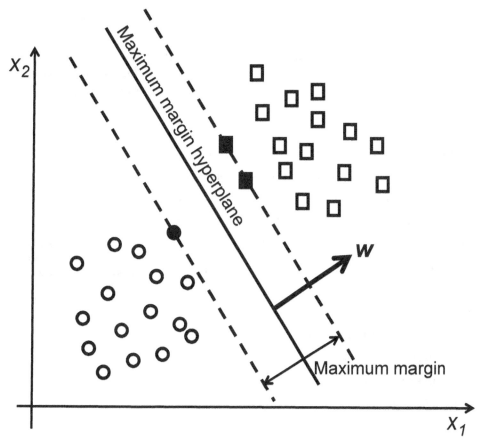

X_2

Maximum margin hyperplane

w

Maximum margin

X_1

Figure 9.3: A support vector machine finds the optimal boundary by determining the maximum margin hyperplane. The weight vector **w** determines the orientation of the boundary and the support vectors (marked in black) define the maximum margin.

classification boundary. Other points, particularly those that lie farther away, do not contribute in defining the boundary. In that respect, only the support vectors have a direct effect on the location of the classification boundary. We can get a further understanding of the support vectors by considering an analogy with static equilibrium in mechanics: The forces $\mathbf{F}_n$ exerted on a static still slab (the decision boundary) defined by the normal vector **w** must satisfy the following equilibrium conditions:

The support vectors are the data points that are the most difficult to classify.

$$\sum_n \mathbf{F}_n = 0, \tag{9.3}$$

These are the equations of mechanical equilibrium.

$$\sum_n \mathbf{s}_n \times \mathbf{F_n} = 0, \tag{9.4}$$

where $\mathbf{s}_n$ are the support vectors in our analogy. Equation (9.4) implies that the total torque exerted on the boundary is zero, and we can think of these vectors as providing support for the hyperplane to be in "static equilibrium". The SVM algorithm was proposed by Vapnik and collaborators[2] in 1992 and variants and its applications are quite varied.

[2] Boser, B. E., I. M. Guyon, and V. N. Vapnik (1992). A training algorithm for optimal margin classifiers. In 5th Annual ACM Workshop on COLT, Pittsburgh, PA, pp. 144–152. ACM Press

Remember that we are interested in obtaining a linear classifier where the margin is as large as possible. Furthermore, let us recall that the distance from a point with coordinates (x_0, y_0) to a line defined by $Ax + By + C = 0$ is given by:

$$d = \frac{|Ax_0 + By_0 + C|}{\sqrt{A^2 + B^2}}, \tag{9.5}$$

and hence the distance from one of the dotted lines in Figure 9.3 and the classification boundary is given by:

$$\frac{|\mathbf{w}^T\mathbf{x} + b|}{||\mathbf{w}||} = \frac{1}{||\mathbf{w}||}, \tag{9.6}$$

This is the distance from the support vector to the classification boundary.

and the margin is therefore twice this amount. It is clear that in order to maximise the margin we need to minimise $||\mathbf{w}||$ with the condition that there are no other points inside the margin. It is possible to express this optimisation problem as:

$$\min_{\mathbf{w},b} \frac{1}{2}||\mathbf{w}||^2, \tag{9.7}$$

$$s.t.: \quad y_i\left(\mathbf{w}^T\mathbf{x}+b\right) \geq 1, \quad i=1,\dots,n.$$

The SVM algorithm as an optimisation problem.

Up to this point we have considered a situation where the data points are linearly separable as depicted in panel a) of Figure 9.1. Nonetheless, in general the data may not necessarily be so nicely separated. In those cases we can consider using a larger margin than the one originally obtained at the expense of incurring some training error. This can be achieved by introducing *slack variables* $\xi_i \geq 0$ and replacing the constraint of the optimisation problem in Equation (9.7) with the following one:

In some cases we can tolerate a larger margin at the expense of incurring some training error. This can be done with the help of slack variables.

$$y_i\left(\mathbf{w}^T\mathbf{x}+b\right) \geq 1-\xi_i, \quad i=1,\dots,n. \tag{9.8}$$

The slack variables generalise our optimisation problem allowing for some misclassification in the training records at a cost C. A data point is misclassified if the value of its slack variable is greater than 1. A bound on the number of misclassified examples is given by the sum of the slack variables. Our optimisation problem can therefore be expressed as:

The slack variables provide us with a way to regularise our optimisation problem.

$$\min_{\mathbf{w},b} \frac{1}{2}||\mathbf{w}||^2 + C\sum\xi_i, \tag{9.9}$$

$$s.t.: \quad y_i\left(\mathbf{w}^T\mathbf{x}+b\right) \geq 1-\xi_i, \quad \xi_i \geq 0.$$

The soft-margin SVM formulation.

This formulation of the support vector machine algorithm is called the **soft-margin SVM**[3]. This is a form of the variance-bias trade-off we know and love. Every constraint imposed can be satisfied for a sufficiently large ξ_i. The constant C is a regularisation parameter: For small values of C we ignore the constraint and we obtain a large margin, whereas for large values of C the constraint is imposed and we end up with a narrow margin. If $C \rightarrow \infty$ then all constraints are imposed and we get a "hard" margin.

[3] Cortes, C. and V. Vapnik (1995). Support vector networks. *Machine Learning 20*, 273–297

Let us start by solving the quadratic programming problem given in Equation (9.7). Since it is a constrained optimisation problem we can start by using the well-known method of Lagrange multipliers[4] where we are interested in optimising a function $f(x,y)$ subject to the constraint $g(x,y) = 0$. We construct a new function called the Lagrangian such that:

[4] Bertsekas, D. (1996). *Constrained Optimization and Lagrange Multiplier Methods*. Athena scientific series in optimization and neural computation. Athena Scientific

$$L(x,y,\alpha) = f(x,y) - \alpha g(x,y), \qquad (9.10)$$

where the new variable α is called the Lagrange multiplier.

The method can be extended to be applied to n dimensions and m constraints. This means that taking the appropriate derivatives to calculate the gradient of the Lagrangian results in $n + m$ equations all set to zero. This optimisation problem can be tackled with the so-called steepest descent algorithm.

The Lagrange multipliers method can be used to solve our optimisation problem.

For the case of our interest we have that $f(\cdot) = \frac{1}{2}||\mathbf{w}||^2$ and $g(\cdot) = y_i\left(\mathbf{w}^T\mathbf{x} + b\right) - 1 = 0$. Our new unconstrained problem in terms of the Lagrangian is expressed as:

$$\min_{\mathbf{w},b} L = \frac{1}{2}||\mathbf{w}||^2 - \sum_{i=1}^{l} \alpha_i y_i \left(\mathbf{w}^T x_i + b\right) + \sum_{i=1}^{l} \alpha_i, \qquad (9.11)$$

The constrained form of the SVM optimisation problem.

where l is the number of training points. We know that the partial derivatives with respect to $\mathbf{w}$ and b should be zero at the minimum of the Lagrangian function. We can then write the following:

$$\partial_{\mathbf{w}} L = \mathbf{w} - \sum_{i=1}^{l} \alpha_i y_i x_i = 0, \qquad (9.12)$$

The partial derivatives of the Lagrangian formulation with respect to $\mathbf{w}$ and b.

$$\partial_b L = \sum_{i=1}^{l} \alpha_i y_i = 0, \qquad (9.13)$$

From the equations above we obtain the following conditions that enable us to find the margin for our SVM problem:

$$\mathbf{w} = \sum_{i=1}^{l} \alpha_i y_i x_i, \qquad (9.14)$$

These are the conditions that enable us to find the margin for our SVM problem.

$$\sum_{i=1}^{l} \alpha_i y_i = 0. \qquad (9.15)$$

The first condition above already tells us something about what the weights $\mathbf{w}$ are: They turn out to be linear combinations of the training inputs and outputs, as well as the values α_i. In turn, we expect that most of the α_i parameters are zero and those that are not will correspond to the actual support vectors.

The weights $\mathbf{w}$ are linear combinations of the training inputs and outputs.

The problem cast above is known as the **primal** form and we can solve it directly in that representation. However, there are advantages in actually solving the **dual** representation of the problem. In that manner, instead of finding the minimum over $\mathbf{w}$ and b, with constraints over the parameters α_i, we can maximise over α subject to conditions (9.14) and (9.15). By substituting these conditions into expression (9.11) we free ourselves from dependencies on $\mathbf{w}$ and b. The Lagrangian in the dual form is given by:

Instead of solving the primal problem we would like to solve the dual one.

In this context α is called the dual variable.

$$\max_{\alpha_i} L_D = -\frac{1}{2} \sum_{i=1}^{l} \sum_{j=1}^{l} \alpha_i \alpha_j y_i y_j \left(\mathbf{x}_i \cdot \mathbf{x}_j \right) + \sum_{i=1}^{l} \alpha_i, \text{ (9.16)}$$

$$\text{subject to } \sum_{i=0}^{l} \alpha_i y_i = 0, \text{ and } a_i \geq 0.$$

The Lagrangian in the dual representation.

Transforming the problem into its dual form may seem a bit over the top, after all we could find the minimum in the primal representation. Nonetheless, there is a substantial gain by using the dual, namely we can solve our problem by performing simple inner products of two vectors and mapping them into the real line $\mathbb{R}$. This is a very important result and we will address it more generally in Section 9.1.2.

The inner product maps two vectors in feature space K into the real line $\mathbb{R}$.

Solving the optimisation problem in Equation (9.16) provides us with the values for the parameters α_i. We can then find the weights $\mathbf{w}$ with the aid of condition (9.14), whereas b can be obtained from a support vector such that $y_i = 1$, giving us the maximal margin hyperplane we were looking for. Finally, we can classify an unseen data point with features $\mathbf{x}$ by looking at the sign of $f(\mathbf{x})$.

Given the dual variables α_i we can calculate $\mathbf{w}$ and b directly.

It is important to note that the vast majority of the weights will have α_i values equal to zero and that only the actual support vectors will survive. This is effectively a way of reducing the dimensionality of our problem. This reduction has been achieved in a more straightforward manner thanks to the application of the inner products in the dual formulation of our problem. In order to provide some intuition about the role of the inner product, we can appeal to the discussions we had in Section 3.8.

> The vast majority of the weights will have α_i values equal to zero, leaving only the support vectors.

The inner product can be used as a measure of the similarity between two vectors defined over an N-dimensional feature space. In a 2D space for example, if two vectors are parallel then their inner product is 1 and we say that the vectors are *completely similar*. If the vectors are perpendicular, then their inner product is zero and we say that they are *completely dissimilar*. This notion of similarity can be extended to higher-dimensional spaces without loss of generality.

> This is the idea behind cosine similarity discussed in Section 3.8.

With the above discussion in mind, in the case of Equation (9.16), if two vectors $\mathbf{x}_i$ and $\mathbf{x}_j$ are dissimilar they will not contribute to the value of L_D. However, if the two vectors are similar we have two possible results. On the one hand, they can predict the same target value $y_i = \pm 1$. In this case the value given by Equation (9.16) will be positive. Remember that we are trying to maximise L_D, and the minus sign attached to the first term of Equation (9.16) implies that the situation described above will decrease the value of the overall expression for L_D. This means that the algorithm reduces the importance of similar vectors that make the same prediction.

> If two vectors are completely dissimilar, they do not contribute to the value of L_D.

> The first term of Equation (9.16) contains the inner product.

On the other hand, if the vectors in question make opposite predictions about the target value y_i, but they are nonetheless similar, then the term containing the inner product is negative. This means that its contribution will increase the value of L_D. This situation helps maximising our objective function. Vectors like these are actually the examples we are interested in, as they are the ones of utmost importance to be able to discriminate our two classes.

One predicts $+1$ whereas the other one -1.

These are the type of vectors we are interested in as they help maximise the value of L_D.

9.1.2   The Kernel Trick

OUR DISCUSSION OF THE SUPPORT vector machine algorithm has relied on the fact that the classes in our problem can be distinguished thanks to a linear boundary. We have seen how the constraint can be relaxed by including slack variables, however the linear separability limitation remains. Although linear separability is a case of interest, we cannot leave aside the fact that there are many cases where a nonlinear boundary exists.

The application of the SVM algorithm is not confined to linearly separable classes.

In those cases we need the implementation of nonlinear support vector machines. The main idea is to obtain a linear boundary by mapping the data into a higher-dimensional space. That is indeed possible but we can come across a few issues. We may not be certain about the type of transformation that needs to be done to obtain linear separability. Even if we did know, the transformation might turn out to be computationally difficult and time-consuming.

The main aim is to obtain a linear boundary by mapping data into a higher-dimensional space.

To a certain extent that is the issue we have already circumvented in the previous section by avoiding the optimisation problem in the primal form shown in Equation (9.11). Instead we used the dual formulation where the inner product enabled us to remap the vectors $\mathbf{x}_i$ into a representation where we did not have to carry out any calculations in the original feature space.

A suitable kernel enables us to carry out the required transformation letting us operate implicitly in the original space.

This is possible thanks to the application of a suitable kernel to carry out a transformation that lets us operate implicitly in the original feature space. This is what is known as **the kernel trick**.

A kernel is a function $K(x, y)$ whose arguments x and y can be real numbers, vectors, functions, etc. It is effectively a map between these arguments and a real value. The operation is independent of the order of the arguments. We are familiar with at least one such kernel: The well-known vector product. With $\mathbf{x}$ and $\mathbf{y}$ being two N-dimensional vectors, the inner product is given by:

This means that $K(x, y) = K(y, x)$.

$$K(\mathbf{x}, \mathbf{y}) = \mathbf{x}^T \mathbf{y} = \sum_{i=1}^{N} x_i y_i. \tag{9.17}$$

The inner product is a well-known kernel.

The kernel trick is a direct implementation of the Mercer theorem: With K being a mapping as defined above and a non-negative definite, symmetric continuous function, there exists a set of functions $\{\phi_i\}$ and a set of positive real numbers $\{\lambda_i\}$ with $i \in \mathbb{N}$ such that

$$K(u, v) = \sum_{i=1}^{\infty} \lambda_i \phi_i(u) \phi_i(v). \tag{9.18}$$

The Mercer theorem is the mathematics behind the kernel trick.

The Mercer theorem can be seen as an analogue of the singular value decomposition we discussed in Section 8.3. In this case the kernel lives in an infinite-dimensional space. So, for a positive-definite symmetric matrix $\mathbf{A}$ we can define a linear operation such that when applied to a vector $\mathbf{x}$, it generates another vector $\mathbf{y}$:

The Mercer theorem can be seen as an analogue of the singular value decomposition.

$$\mathbf{A}\mathbf{x} = \mathbf{y}, \quad \text{or} \quad \sum_{m=1}^{N} a[m,n]x[m] = y[m]. \tag{9.19}$$

This is the usual matrix multiplication that we know and love.

The eigenvalues and corresponding eigenfunctions are defined as:

$$\sum_{m=1}^{N} a[m,n]\phi_i[m] = \lambda_i\phi_i[n]. \tag{9.20}$$

The eigenvalues are non-negative and the eigenfunctions are orthonormal.

The eigenvalues are non-negative and the eigenfunctions are orthonormal. As such, the eigenfunctions that correspond to non-zero eigenvalues form a basis for the matrix $\mathbf{A}$, and we can therefore decompose it as:

$$a[m,n] = \sum_{i=1}^{N} \lambda_i\phi_i[m]\phi_j[n]. \tag{9.21}$$

We can decompose the matrix in this way.

The kernel trick implies that we do not need to compute, or even know, the functions ϕ_i. Instead the kernel defines the appropriate inner products in the transformed space. In that way, it is possible to define nonlinear support vector machines where the objective function can be written as:

The kernel trick implies that we do not need to compute, or even know, the functions ϕ_i.

$$L_D = \sum_{i=1}^{l} \alpha_i - \frac{1}{2}\sum_{i=1}^{l}\sum_{j=1}^{l} \alpha_i\alpha_j y_i y_j K(x_i, x_j). \tag{9.22}$$

We have a choice of kernels to use. Some of the more popular ones include:

- Linear: $K(\mathbf{x}, \mathbf{y}) = \mathbf{x}^T \mathbf{y}$

- Polynomial: $K(\mathbf{x}, \mathbf{y}) = \left(\mathbf{x}^T \mathbf{y} + 1\right)^d$

- Gaussian: $K(\mathbf{x}, \mathbf{y}) = \exp\left(-\gamma \|\mathbf{x} - \mathbf{y}\|^2\right)$

- Sigmoid: $K(\mathbf{x}, \mathbf{y}) = \tanh\left(\kappa \mathbf{x} \cdot \mathbf{y} - \delta\right)$

> Here is a selection of some popular kernels that we can use.

We can certainly take our pick, but remember that more complicated models may not lead to good generalisation. Beware of overfitting.

9.1.3   SVM in Action: Regression

THE SUPPORT VECTOR MACHINE ALGORITHM can be applied in regression problems as a way to optimise the generalisation boundaries for the regression line. In this case the feature variables are first mapped onto the higher-dimensional space and then a linear model is used.

> SVM can be used in regression problems.

The generalisation performance of the SVM depends on the kernel used, but as with other regularised models, it also depends on the hyperparameter C introduced in Equation (9.9). This hyperparameter gives us the chance to fine-tune the model complexity: If C tends to infinity, then we are effectively optimising the model given only the data observed, and disregarding completely the complexity of the model.

> Remember, it is all about the bias and variance trade-off

Let us see how we can run SVM for regression using the mammals data we introduced in Chapter 4. Let us recall that the dataset looks at the relationship of the body mass of an animal and the mass of its brain. As before, let us start by importing some useful Python libraries:

The data is available at `http://dx.doi.org/10.6084/m9.figshare.1565651` as well as `http://www.statsci.org/data/general/sleep.html`.

```
%pylab inline
import numpy as np
import matplotlib.pyplot as plt
import pandas as pd
```

We will now load the dataset into a Pandas dataframe and sort it. We do this so that at a later stage our plot looks as we expect it and not like a bad piece of pseudo-modern art:

```
mammals = pd.read_csv(u'./Data/mammals.csv')\
    .sort_values('body')
```

We sort the data with the method `.sort_values()`.

Let us load the values of the required columns into appropriate variables for ease of manipulation:

```
body = mammals[ ['body'] ].values
brain = mammals['brain'].values
```

Please note that we have also created objects to hold the values of the quantities of interest. We are interested in comparing the result obtained using the well-known linear regression models we discussed in Chapter 4 with the SVM algorithm. Let us load the relevant models from Scikit-learn:

We will compare the results of the SVM with the simple linear regression model.

```
from sklearn.linear_model import LinearRegression
from sklearn import svm
```

SVM contains methods for Support Vector Regression
including SVR and LinearSVR. The former accepts as input
different kernels and the latter is similar to SVR with
kernel='linear'. LinearSVR has more flexibility in the
choice of penalties and loss functions. It is also worth
mentioning that the default kernel for SVR is a Gaussian
kernel, also known as a radial basis function (RBF) kernel.
Let us instantiate a couple of SVR models with a fixed value
for the hyperparameter C:

Scikit-learn has the following
implementations for Support
Vector Regression: SVR and
LinearSVR.

```
svm_lm = svm.SVR(kernel='linear', C=1e1)
svm_rbf = svm.SVR(kernel='rbf', C=1e1)
```

We instantiate a model with a
linear kernel and one with a
Gaussian one.

The first model above is a SVM for regression with a linear
kernel, where as the second one has a Gaussian kernel. We
can now train our models:

```
svm_lm.fit(np.log(body), np.log(brain))
svm_rbf.fit(np.log(body), np.log(brain))
```

We train our models with the log
of the available variables.

For comparison, we will now instantiate and fit a regression
model with logarithmic transformation:

```
logfit = LinearRegression().fit(np.log(body),\
np.log(brain))
```

Please note that we can chain the
methods.

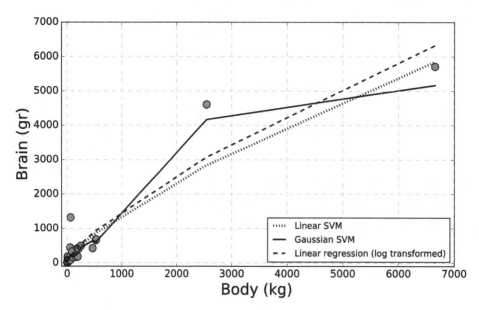

Figure 9.4: A comparison of the regression curves obtained using a linear model, and two SVM algorithms: one with a linear kernel and the other one with a Gaussian one.

The predictions of the models above can be easily obtained with the `predict` method for each of the models. In this case we are directly attaching them to the Pandas dataframe we started with:

```
mammals['log_regr'] = np.exp(logfit.\
predict(np.log(body)))


mammals['linear_svm'] = np.exp(svm_lm.\
predict(np.log(body)))


mammals['rbf_svm'] = np.exp(svm_rbf.\
predict(np.log(body)))
```

We can obtain the predictions for all the trained models.

The result of the three regression procedures applied to the mammals dataset can be seen in Figure 9.5. Note that in this case the linear kernel performs no better and even worse than the simple linear model we saw in Chapter 4. The Gaussian kernel did not require us to make any explicit transformations and, in this case, gets closer to the values observed in the original dataset. Also note that close to the origin the Gaussian kernel produced some wiggles in the regression curve. Finally, please remember that overfitting is still an adversary that needs to be considered. We leave the tuning of the hyperparameter C as well as the implementation of cross-validation as an exercise for the reader.

The Gaussian kernel seems to give better results than the other two models.

9.1.4   SVM in Action: Classification

THE SECOND APPLICATION OF SUPPORT vector machines we will see is that of classification. As we saw in Figure 9.3, the maximum margin hyperplane serves as a boundary between different classes. See Equation (9.2) for the mathematical expression of this statement.

Let us implement the SVM for classification using the wine data we encountered in Chapter 5. The dataset can be found in the UCI Machine Learning Repository under "Wine Dataset"[5] and is available at `http://archive.ics.uci.edu/ml/datasets/Wine`. Recall that the data records the results of chemical analysis of Italian wines grown in the same region from three different cultivars.

[5] Lichman, M. (2013a). UCI Machine Learning Repository, Wine Data. `https://archive.ics.uci.edu/ml/datasets/Wine`. University of California, Irvine, School of Information and Computer Sciences

Let us start by loading the appropriate libraries into Python so that we can read the CSV file where the data is contained:

```
%pylab inline
import numpy as np
import pandas as pd
import matplotlib.pyplot as plt
```

Using Pandas, we will load the data into a dataframe called wine. The data is such that the target variable is contained in the same table and thus we separate the Cultivar feature into a target variable Y, leaving the rest of the features in the variable X:

The data is loaded into Pandas, and the target variable is separated from the rest of the dataset.

```
wine = pd.read_csv(u'./Data/wine.csv')

X = wine.drop(['Cultivar'], axis=1).values
Y = wine['Cultivar'].values
```

In the same way that we dealt with this dataset for clustering in Section 5.2.2, we will only use the Alcohol and Colour Intensity features in our classification task. This will make our example simpler and easier to understand:

We will only use the Alcohol and Colour Intensity features.

```
X1=wine[['Alcohol','Colour_Intensity']].values
```

We can now split our data set into training and testing for cross-validation purposes:

```
import sklearn.model_selection as ms

XTrain, XTest, YTrain, YTest =\
ms.train_test_split(X1, Y,\
test_size= 0.3, random_state=7)
```

As good jackalope datascientists, we are well-acquainted with creating training and testing datasets.

Scikit-learn provides implementations of Support Vector Classification including SVC and LinearSVC. As with the Support Vector Regression case, SVC takes as input different kernels, whereas LinearSVC is similar to SVC with kernel='linear'. Once again, as in the regression case, LinearSVC has more flexibility in the choice of penalties and loss functions. Finally, remember that the default kernel for SVC is a Gaussian one or RBF.

For classification, Scikit-learn has SVC and LinearSVC implementations.

Remember that RBF stands for radial basis function.

Let us create an instance of the SVC classifier with a Gaussian kernel as follows:

```
from sklearn import svm
SVMclassifier = svm.SVC()
```

Remember that the default kernel for SVC is a Gaussian one.

We can use GridSearchCV in conjunction with the training set created above to find a suitable value for the hyperparameter C in our model. In this case we set up a dictionary for our search with values ranging between 0.5 and 2:

```
Cval = 2. ** np.arange(-1, 1.2, step=0.2)
n_grid = [{'C':Cval}]
```

We will search for the best hyperparameter C.

We can now set up our cross-validated grid search with a support vector machine for classification:

```
from sklearn.model_selection import GridSearchCV
cv_svc = GridSearchCV(estimator=SVMclassifier,\
param_grid=n_grid,\
cv=ms.KFold(n_splits=100))
```

We use GridSearchCV to find the best value for the hyperparameter. In this case we are using 100 folds.

Let us now apply the search to data coming from the training set we constructed above. We can also store the value of the best value for the parameter C for future use:

```
cv_svc.fit(XTrain, YTrain)

best_c = cv_svc.best_params_['C']
```

The model can finally be fitted.

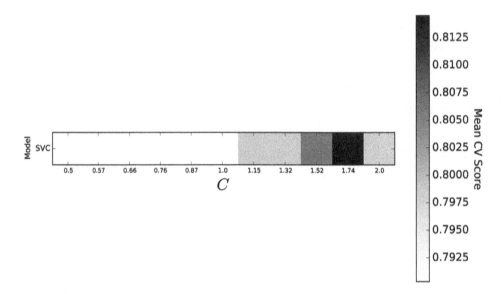

Figure 9.5: Heatmap of the mean cross-validation scores for the a support vector machine algorithm with a Gaussian kernel for different values of the parameter C.

Let us see what the best parameter found is in this case:

```
> print(''The best parameter is: C ='', best_c)

The best parameters is: C = 1.74110112659
```

In Figure 9.5 we can see a heatmap of the search performed with the different values for C. As we can see, for $C = 1.7411$ we find the best mean cross-validation score for the model. With this value it is now possible to construct a model to be used for training and testing:

```
svc_clf = svm.SVC(C=best_c)
svc_clf.fit(XTrain, YTrain)
```

We now create a model using the best parameter found and train it.

It is now possible to obtain the predictions provided by the model for the data points in the testing set. Let us take a look:

```
y_p = svc_clf.predict(XTest)
```

We obtain the predictions for the testing dataset.

For completeness, let us take at look ar the classification report obtained for the testing dataset:

```
> from sklearn import metrics
> print(metrics.classification_report(y_p, YTest))


            precision    recall  f1-score   support
         1       0.85      0.69      0.76        16
         2       0.92      0.92      0.92        24
         3       0.76      0.93      0.84        14


avg / total    0.86      0.85      0.85        54
```

The classification report gives us information about the precision and recall among other things.

Using the same value for C obtained above, let us compare the classification boundaries obtained using different kernels:

```
C = best_c


svc = svm.SVC(kernel='linear', C=C).\
            fit(XTrain, YTrain)


rbf_svc = svm.SVC(kernel='rbf', gamma=0.7,\
            C=C).fit(XTrain, YTrain)


poly_svc = svm.SVC(kernel='poly', degree=3,\
            C=C).fit(XTrain, YTrain)


lin_svc = svm.LinearSVC(C=C).fit(XTrain, YTrain)
```

We are creating support vector machine models with linear, Gaussian and polynomial kernels for comparison.

First, we are building a support vector machine (SVM) with a linear kernel with the SVC implementation. A second model has a Gaussian kernel with gamma=0.7 and a third one with a cubic polynomial kernel (degree=3). Finally, with LinearSVC we create a model using an alternative linear kernel implementation. A comparison of these models is shown Figure 9.6. We can see how the boundaries created by the linear kernels are given by straight lines, whereas the nonlinear ones provide us with more intricate boundaries.

Note that the boundaries for linear kernels are lines, whereas the ones for nonlinear kernels are more intricate.

The plots in Figure 9.6 also show the data points in the testing set coloured by the class to which they belong. We can see the distribution of misclassified points given by each of the four models used. It is clear that we can use more features than the two chosen in this example. This will make the visualisation step more challenging than the 2D case shown here.

SVM with linear kernel

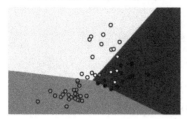

SVM with RBF kernel

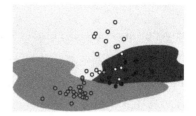

SVM with polynomial (degree 3) kernel

LinearSVM (linear kernel)

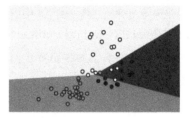

Figure 9.6: A comparison of the classification boundaries obtained using support vector machine algorithms with different implementations: SVC with a linear, Gaussian and degree-3 polynomial kernels, and LinearSVC.

9.2   *Summary*

THE KERNEL TRICK IS A useful tool to consider alongside other techniques such as feature selection and dimensionality reduction to tackle any machine learning problem that we encounter.

In this chapter we saw how appropriate mappings into higher-dimensional feature spaces can be used to make regression or classification possible without actually having to carry out explicitly any calculations in the high-dimensional space. This is possible thanks to the so-called *kernel trick*. We applied this trick in the implementation of the Support Vector Machine (SVM) algorithm.

A support vector machine is a binary linear classifier were the classification boundary is constructed so as to minimise the generalisation error in our learning task. The main idea behind SVM is that data may be linearly separable in the high-dimensional space, but not linearly separable in the original feature space. The use of different kernels provides us with the flexibility to find nonlinear boundaries that separate the different classes in our problem as demonstrated in this chapter.

Pipelines in Scikit-Learn

THE TERM PIPELINE IS USED to describe a series of ordered concatenated data transformations and manipulations. The order of the transformations is important as each phase of a pipeline feeds from the previous one: The outcome of any given step serves as the input for the next one as data flows through the pipeline from beginning to end.

A pipeline is a series of ordered concatenated data transformations and manipulations.

Data pipelines are very useful as they are effectively automated transformations used to perform routine data maintenance and analysis tasks, ensuring the validity of the output data to be fed to the next stage of the workflow. For instance in a required workflow it may be necessary to ensure that the data has the correct units and is scaled before imputing any missing values and is ready to be used for training a particular algorithm.

They help us manage our workflow by defining the steps we need to take.

It is possible to implement pipelines in Scikit-learn, helping us improve our code and manage our models. A pipeline can be used to amalgamate all the steps we may need to prepare our data and make appropriate predictions. In Section 8.2.2 we applied a pipeline to reduce the

We discussed an example in Section 8.2.2.

dimensionality of the Iris dataset with principal component analysis and then use the result in a logistic regression.

Let us see another example implementing a pipeline for a LASSO regression using the Boston housing dataset included in Scikit-learn. The dataset has information from the US Census Service concerning housing in the area of Boston, Massachusetts. The dataset was originally published by Harrison and Rubinfeld[6] and has 13 attributes as follows:

[6] Harrison Jr, D. and Rubinfeld, D. L. (1978). Hedonic housing prices and the demand for clean air. *J. Environ. Economics & Management* 5, 81–102

- CRIM - per capita crime rate by town

- ZN - proportion of residential land zoned for lots over 25,000 sq.ft.

- INDUS - proportion of non-retail business acres per town

- CHAS - Charles River dummy variable (1 if tract bounds river; zero otherwise)

- NOX - nitric oxides concentration (parts per 10 million)

- RM - average number of rooms per dwelling

These are the features in the Boston Housing dataset in Scikit-learn.

- AGE - proportion of owner-occupied units built prior to 1940

- DIS - weighted distances to five Boston employment centres

- RAD - index of accessibility to radial highways

- TAX - full-value property-tax rate per $10,000

- PTRATIO - pupil-teacher ratio by town

- B - $1000(Bk - 0.63)^2$ where Bk is the proportion of black population by town

- LSTAT - percentage lower status of the population

Let us load the dataset:

```
%pylab inline
from sklearn.datasets import load_boston
boston = load_boston()
X = boston[''data'']
Y = boston[''target'']
names = boston[''feature_names'']
```

The Boston Housing dataset can be loaded from `load_boston`.

We will build a pipeline that takes into account two main steps, one to standardise the variables as described in Section 4.6 and then use the result in a LASSO model. We will use the pipeline to search for the appropriate hyperparameter for the model and finally use the parameter found to train the model and score it.

Our pipeline will consist of a standardisation step and a LASSO model.

Let us first load some useful modules: We will use `preprocessing` to standardise our variables, `cv` to create our training and testing partitions, `Lasso` for modelling our data, `GridSearchCV` to search for the optimal hyperparameter and `Pipeline` to construct our pipeline.

```
from sklearn import preprocessing
import sklearn.model_selection as ms
from sklearn.linear_model import Lasso
from sklearn.model_selection import GridSearchCV
from sklearn.pipeline import Pipeline
```

These are the modules and methods we will use in this analysis.

We need to create instances for the steps that will be included in our pipeline. In this case standardisation of our variables with Scikit-learn's `StandardScaler` and a LASSO model.

```
std_scaler = preprocessing.StandardScaler()
lasso1 = Lasso()
```

We need to instantiate the steps to be used in our pipeline.

We can now define our pipeline. We have to provide a label to each of our steps so that we can refer to them later on in the process. Let us call the standardisation step `scaling` and the model `mylasso`:

```
pipe = Pipeline(steps=[('scaling', std_scaler ),\
('mylasso', lasso1)])
```

Note that we are providing a label to each of the steps in our pipeline.

We need to partition our data into training and testing:

```
XTrain, XTest, yTrain, yTest =\
ms.train_test_split(X, Y,\
test_size=0.2, random_state=1)
```

As usual we partition our data into training and testing.

Let us define a set of values to search for the hyperparameter.

```
lambda_range = linspace(0.001, 0.5, 250)
```

We will perform a search over a range of values for the model's hyperparameter.

We can now pass our pipeline to the exhaustive search module and fit the search with the training data as follows:

```
cv_lasso = GridSearchCV(pipe,\
dict(mylasso__alpha=lambda_range),\
cv=ms.KFold(n_splits=100))


cv_lasso.fit(XTrain,yTrain)
```

We pass our pipeline as the model to be used by `GridSearchCV`. We then execute the search over the training dataset.

The grid search will take the raw training data and put it through the pipe: First it calls the standardisation process and the result will be fed to the LASSO model for each of the hyperparameter values for cross validation. We can take a look at the best parameter obtained from this process as follows:

```
bestLambda=cv_lasso.best_params_['mylasso__alpha']
print(bestLambda)


0.229457831325
```

The result of the hyperparameter search is prefixed by the label provided when defining our pipeline.

Please note that we are referring to the parameter from the LASSO model by prefixing it with the label given to the pipeline followed by a double underscore.

Now that we have found the optimal hyperparameter we can set it as a parameter in the pipeline with the help of the `set_params` method:

```
pipe.set_params(mylasso__alpha=bestLambda)
```

We can pass the new parameter to the pipe with `set_params`.

We are now in a position to train our model:

```
> BostonLassoModel = pipe.fit(XTrain, yTrain)
```

The pipeline can be directly used to train our final model.

Let us take a look at the coefficients we obtained:

```
> BostonLassoModel.named_steps['mylasso'].coef_

array([-0.3716588 ,  0.43517511,
        -0.        ,  0.47183528,
        -1.05678543,  2.4463162 ,
        -0.        , -1.51971828,
         0.        , -0.        ,
        -1.83958382,  0.44593894,
        -3.83998777])
```

Note that a few of the attributes have coefficients equal to zero.

Since we have used a LASSO model it is no surprise that some of the coefficients used in the model have shrunk down to zero. You can take a look at the details in Section 4.9.

Finally, let us score the model and create predictions for the testing partition:

```
> BostonLassoModel.score(XTest, yTest)

0.73531414540197193

> Boston_Pred = BostonLassoModel.predict(XTest)
```

Finally we can score and obtain predictions from our model. Et voilà!

Bibliography

Allison, T. and D. V. Cicchetti (1976, Nov 12). Sleep in mammals: ecological and constitutional correlates. *Science 194*, 732–734.

Bayes, T. (1763). An essay towards solving a problem in the doctrine of chances. *Philosophical Transactions 53*, 370–418.

Bellman, R. (1961). *Adaptive Control Processes: A Guided Tour*. Rand Corporation. Research studies. Princeton U.P.

Bertsekas, D. (1996). *Constrained Optimization and Lagrange Multiplier Methods*. Athena scientific series in optimization and neural computation. Athena Scientific.

Borges, J. L. (1984). *El Libro de Arena*. El Ave Fénix. Plaza & Janés.

Boser, B. E., I. M. Guyon, and V. N. Vapnik (1992). A training algorithm for optimal margin classifiers. In *5th Annual ACM Workshop on COLT*, Pittsburgh, PA, pp. 144–152. ACM Press.

Breiman, L. (1996). Bagging predictors. *Machine Learning 24*(2), 123–140.

Breiman, L. (2001). Random forests. *Machine Learning* 45(1), 5–32.

Cole, S. (2004). History of fingerprint pattern recognition. In N. Ratha and R. Bolle (Eds.), *Automatic Fingerprint Recognition Systems*, pp. 1–25. Springer New York.

Continuum Analytics (2014). Anaconda 2.1.0. `https://store.continuum.io/cshop/anaconda/`.

Cortes, C. and V. Vapnik (1995). Support vector networks. *Machine Learning* 20, 273–297.

Cover, T. M. (1969). Nearest neighbor pattern classification. *IEEE Trans. Inform. Theory IT-13*, 21–27.

Devlin, K. (2010). *The Unfinished Game: Pascal, Fermat, and the Seventeenth-Century Letter That Made the World Modern*. Basic ideas. Basic Books.

DLMF (2015). NIST Digital Library of Mathematical Functions. http://dlmf.nist.gov/, Release 1.0.10 of 2015-08-07.

Downey, A. (2012). *Think Python*. O'Reilly Media.

Duffy, F. H. et al. Unrestricted principal components analysis of brain electrical activity: Issues of data dimensionality, artifact, and utility. *Brain Topography* 4(4), 291–307.

Eysenck, M. and M. Keane (2000). *Cognitive Psychology: A Student's Handbook*. Psychology Press.

Farris, J. S. (1969). On the cophenetic correlation coefficient. *Systematic Biology* 18(3), 279–285.

Fawcett, T. (2006). An introduction to ROC analysis. *Patt. Recog. Lett. 27,* 861–874.

Fisher, R. A. (1936). The use of multiple measurements in taxonomic problems. *Annals of Eugenics 7*(2), 179–188.

Fold-it. Solve puzzles for science. `https://fold.it/portal/`.

Freedman, D., R. Pisani, and R. Purves (2007). *Statistics.* International student edition. W.W. Norton & Company.

Freund, Y. and R. Schapire (1997). A decision-theoretic generalization of on-line learning and an application to boosting. *J. Comp. and Sys. Sciences 55*(1), 119–139.

Galati, G. (2015). *100 Years of Radar.* Springer International Publishing.

Galton, F. (1886). Regression Towards Mediocrity in Hereditary Stature. *The Journal of the Anthropological Institute of Great Britain and Ireland 15,* 246–263.

Galton, F. (1907). Vox populi. *Nature 75*(1949), 450–451.

Geurts, P., D. Ernst, and L. Wehenkel (2006). Extremely randomized trees. *Machine Learning 63,* 3–42.

Gilder, J. and A. Gilder (2005). *Heavenly Intrigue: Johannes Kepler, Tycho Brahe, and the Murder Behind One of History's Greatest Scientific Discoveries.* Knopf Doubleday Publishing Group.

Golub, G. and C. Van Loan (2013). *Matrix Computations.* Johns Hopkins Studies in the Mathematical Sciences. Johns Hopkins University Press.

Harrison Jr, D. and Rubinfeld, D. L. (1978). Hedonic housing prices and the demand for clean air. *J. Environ. Economics & Management 5*, 81–102.

Hilbert, D. (1904). Grundzüge einer allgeminen Theorie der linaren Integralrechnungen. (Erste Mitteilung). *Nachrichten von der Gesellschaft der Wissenschaften zu Göttingen, Mathematisch-Physikalische Klasse*, 49–91.

Hoerl, A. E. and R. W. Kennard (1970). Ridge regression: Biased estimation for nonorthogonal problems. *Technometrics 12*(3), 55–67.

Hu, Y., Y. Koren, and C. Volinsky (2008). Collaborative filtering for implicit feedback datasets. In *Proceedings of the 2008 Eighth IEEE International Conference on Data Mining*, ICDM '08, Washington, DC, USA, pp. 263–272. IEEE Computer Society.

Hunt, E. B., J. Marin, and P. J. Stone (1966). *Experiments in induction*. New York: Academic Press.

Kaggle (2012). Titanic: Machine Learning from Disaster. https://www.kaggle.com/c/titanic.

Langtangen, H. (2014). *A Primer on Scientific Programming with Python*. Texts in Computational Science and Engineering. Springer Berlin Heidelberg.

Laplace, P. and A. Dale (2012). *Pierre-Simon Laplace Philosophical Essay on Probabilities: Translated from the fifth French edition of 1825 With Notes by the Translator*. Sources in the History of Mathematics and Physical Sciences. Springer New York.

Le, Q. V., R. Monga, M. Devin, G. Corrado, K. Chen, M. Ranzato, J. Dean, and A. Y. Ng (2011). Building high-level features using large scale unsupervised learning. *CoRR abs/1112.6209*.

Lehren, A. W. and Baker, A. (2009, Jun 18th). In New York, Number of Killings Rises With Heat. *The New York Times*.

Lichman, M. (2013a). UCI Machine Learning Repository, Wine Data. `https://archive.ics.uci.edu/ml/datasets/` `Wine`. University of California, Irvine, School of Information and Computer Sciences.

Lichman, M. (2013b). UCI Machine Learning Repository, Wisconsin Breast Cancer Database. `https://archive.` `ics.uci.edu/ml/datasets/Breast+Cancer+Wisconsin+` `(Original)`. University of California, Irvine, School of Information and Computer Sciences.

Lima, M. (2011). *Visual Complexity: Mapping Patterns of Information*. Princeton Architectural Press.

Lima, M. and B. Shneiderman (2014). *The Book of Trees: Visualizing Branches of Knowledge*. Princeton Architectural Press.

Lohr, S. (2014, Aug 17th). For Big-Data Scientists, 'Janitor Work' Is Key Hurdle to Insights. *The New York Times*.

MacQueen, J. (1967). Some Methods for classification and Analysis of Multivariate Observations. In *Proceedings of 5-th Berkeley Symposium on Mathematical Statistics and Probability*. University of California Press.

Mangasarian, O. L. and W. H. Wolberg (1990, Sep.). Cancer diagnosis via linear programming. *SIAM News* 25(5), 1 & 18.

Martin, D. (2003, Jan 19th). Douglas Herrick, 82, Dies; Father of West's Jackalope. *The New York Times*.

McCandless, D. (2009). *Information is Beautiful*. Collins.

McGrayne, S. (2011). *The Theory that Would Not Die: How Bayes' Rule Cracked the Enigma Code, Hunted Down Russian Submarines, & Emerged Triumphant from Two Centuries of Controversy*. Yale University Press.

McKinney, W. (2012). *Python for Data Analysis: Data Wrangling with Pandas, NumPy, and IPython*. O'Reilly Media.

Milligan, Glenn W. and Cooper, Martha C. (1988). A study of standardization of variables in cluster analysis. *Journal of Classification* 5(2), 181–204.

Pearson, K (1904). On the theory of contingency and its relation to association and normal correlation. In *Mathematical Contributions to the Theory of Evolution*. London, UK: Dulau and Co.

Pedregosa, F., G. Varoquaux, A. Gramfort, V. Michel, et al. (2011). Scikit-learn: Machine learning in Python. *Journal of Machine Learning Research* 12, 2825–2830.

Python Software Foundation (1995). Python reference manual. http://www.python.org.

R Core Team (2014). R: A language and environment for statistical computing. http://www.R-project.org.

Rogel-Salazar, J. (2014). *Essential MATLAB and Octave.*
Taylor & Francis.

Rogel-Salazar, J. (2016a, Jan).   Data Science Tweets.
10.6084/m9.figshare.2062551.v1.

Rogel-Salazar, J. (2016b, Jan).   Jackalope Image.
10.6084/m9.figshare.2067186.v1.

Rogel-Salazar, J. and N. Sapsford (2014). Seasonal effects
in natural gas prices and the impact of the economic
recession. *Wilmott 2014(74),* 74–81.

Rousseeuw, P. J. (1987). Silhouettes: a Graphical Aid to the
Interpretation and Validation of Cluster Analysis. *Comp. and
App. Mathematics 20,* 53–65.

Scientific Computing Tools for Python (2013). NumPy.
http://www.numpy.org.

Takács, G. and D. Tikk (2012). Alternating least squares
for personalized ranking. In *Proceedings of the Sixth ACM
Conference on Recommender Systems,* RecSys '12, New York,
NY, USA, pp. 83–90. ACM.

Tibshirani, R. (1996). Regression Shrinkage and Selection
via the Lasso. *J. R. Statist. Soc. B 58(1),* 267–288.

Toelken, B. (2013). *The Dynamics of Folklore.* University Press
of Colorado.

Töscher, A. and M. Jahrer (2009). The BigChaos solution
to the Netflix grand prize. http://www.netflixprize.com/
assets/GrandPrize2009_BPC_BigChaos.pdf.

Turing, A. M. (1936). On computable numbers, with an application to the Entsheidungsproblem. *Proceedings of the London Mathematical Society 42*(2), 230–265.

Turing, A. M. (1950). Computing machinery and intelligence. *Mind 59*, 433–460.

Weir, A. (2014). *The Martian: A Novel*. Crown/Archetype.

Wolpert, D. H. (1992). Stacked generalization. *Neural Networks 5*(2), 241–259.

Zimmer, C. (2012). *Rabbits with Horns and Other Astounding Viruses*. Chicago Shorts. University of Chicago Press.

Zingg, R., J. Fikes, P. Weigand, and C. de Weigand (2004). *Huichol Mythology*. University of Arizona Press.

Zooniverse. Projects. `https://www.zooniverse.org/projects`.

Index